Salt Taste, Nutrition, and Health

Salt Taste, Nutrition, and Health

Special Issue Editor

Albertino Bigiani

MDPI • Basel • Beijing • Wuhan • Barcelona • Belgrade • Manchester • Tokyo • Cluj • Tianjin

Special Issue Editor
Albertino Bigiani
Dipartimento di Scienze Biomediche, Metaboliche e Neuroscienze
Università di Modena e Reggio Emilia
Italy

Editorial Office
MDPI
St. Alban-Anlage 66
4052 Basel, Switzerland

This is a reprint of articles from the Special Issue published online in the open access journal *Nutrients* (ISSN 2072-6643) (available at: https://www.mdpi.com/journal/nutrients/special_issues/Salt_Taste).

For citation purposes, cite each article independently as indicated on the article page online and as indicated below:

LastName, A.A.; LastName, B.B.; LastName, C.C. Article Title. *Journal Name* **Year**, *Article Number*, Page Range.

ISBN 978-3-03936-465-7 (Hbk)
ISBN 978-3-03936-466-4 (PDF)

Cover image courtesy of Maik Behrens and Kristina Lossow.

Contents

About the Special Issue Editor vii

Albertino Bigiani
Salt Taste, Nutrition, and Health
Reprinted from: *Nutrients* **2020**, *12*, 1537, doi:10.3390/nu12051537 1

Albertino Bigiani
Does ENaC Work as Sodium Taste Receptor in Humans?
Reprinted from: *Nutrients* **2020**, *12*, 1195, doi:10.3390/nu12041195 7

Noriatsu Shigemura, Shingo Takai, Fumie Hirose, Ryusuke Yoshida, Keisuke Sanematsu and Yuzo Ninomiya
Expression of Renin-Angiotensin System Components in the Taste Organ of Mice
Reprinted from: *Nutrients* **2019**, *11*, 2251, doi:10.3390/nu11092251 17

Camilla Cattaneo, Patrizia Riso, Monica Laureati, Giorgio Gargari and Ella Pagliarini
Exploring Associations between Interindividual Differences in Taste Perception, Oral Microbiota Composition, and Reported Food Intake
Reprinted from: *Nutrients* **2019**, *11*, 1167, doi:10.3390/nu11051167 35

Kristina Lossow, Wolfgang Meyerhof and Maik Behrens
Sodium Imbalance in Mice Results Primarily in Compensatory Gene Regulatory Responses in Kidney and Colon, but Not in Taste Tissue
Reprinted from: *Nutrients* **2020**, *12*, 995, doi:10.3390/nu12040995 53

Mee-Ra Rhyu, Ah-Young Song, Eun-Young Kim, Hee-Jin Son, Yiseul Kim, Shobha Mummalaneni, Jie Qian, John R. Grider and Vijay Lyall
Kokumi Taste Active Peptides Modulate Salt and Umami Taste
Reprinted from: *Nutrients* **2020**, *12*, 1198, doi:10.3390/nu12041198 75

Celeste Ferraris, Alexandria Turner, Kiranjit Kaur, Jessica Piper, Martin Veysey, Mark Lucock and Emma L. Beckett
Salt Taste Genotype, Dietary Habits and Biomarkers of Health: No Associations in an Elderly Cohort
Reprinted from: *Nutrients* **2020**, *12*, 1056, doi:10.3390/nu12041056 95

Nikolina Nika Veček, Lana Mucalo, Ružica Dragun, Tanja Miličević, Ajka Pribisalić, Inga Patarčić, Caroline Hayward, Ozren Polašek and Ivana Kolčić
The Association between Salt Taste Perception, Mediterranean Diet and Metabolic Syndrome: A Cross-Sectional Study
Reprinted from: *Nutrients* **2020**, *12*, 1164, doi:10.3390/nu12041164 109

Eun-Gyung Mun, Jung Eun Park and Youn-Soo Cha
Effects of Doenjang, a Traditional Korean Soybean Paste, with High-Salt Diet on Blood Pressure in Sprague–Dawley Rats
Reprinted from: *Nutrients* **2019**, *11*, 2745, doi:10.3390/nu11112745 131

About the Special Issue Editor

Albertino Bigiani (Professor) Degree in Biological Sciences (University of Modena, Modena, Italy); Ph.D. in Neuroscience (University of Pisa, Pisa, Italy); Postdoctoral fellow with Prof. Stephen D. Roper (Department of Anatomy and Neurobiology, Colorado State University, Fort Collins, Colorado, USA). He is now Professor of Physiology and has a tenured position at the University of Modena and Reggio Emilia (Italy). He received training in membrane electrophysiology, and has expertise in the field of cell physiology. His current research interests are the mechanisms used by taste cells to detect sodium ions (Na^+), which are responsible for the taste of table salt (salty). Website: https://www.researchgate.net/profile/Albertino_Bigiani.

nutrients

MDPI

Editorial

Salt Taste, Nutrition, and Health

Albertino Bigiani

Dipartimento di Scienze Biomediche, Metaboliche e Neuroscienze, Università di Modena e Reggio Emilia, 41125 Modena, Italy; albertino.bigiani@unimore.it; Tel.: +39-059-205-5349

Received: 19 May 2020; Accepted: 22 May 2020; Published: 25 May 2020

The sodium ion (Na^+) is essential for life. Na^+ is the main cation in the extracellular fluid bathing all our cells, and it is also a key element in many body secretions. In addition to determining the extracellular fluid osmolality, Na^+ is involved in several physiological processes that would be impaired by its deficiency—without Na^+, neurons and muscle cells would not be able to generate electrical impulses, the intestinal adsorption of nutrients would be undermined, and the kidneys would not work properly. To maintain stable levels of Na^+ in our body, losses through kidneys, the gastrointestinal tract, and sweating have to be balanced by the ingestion of foods containing this mineral. Our ability to recognize Na^+ relies mainly on the activity of the taste system. Usually, we consume sodium in the form of sodium chloride (NaCl), the common table salt. Moderate concentrations of salt induce a specific "salty" sensation that is appealing and appetitive for us. However, saltiness can be perceived as pleasant or unpleasant according to the salt concentration and the medium in which salt occurs. For example, we dislike salt concentrations >150 mM in aqueous solution, but we found them palatable when salt is in foodstuffs. The positive hedonic tone (liking) of salt taste may expose us to salt overconsumption. This is further exacerbated by the widespread use of NaCl as preservative and as flavoring agent in processed foods. When consumed in excess, salt is detrimental to health: in fact, excessive sodium intake is linked to the development of hypertension and related pathologies.

This Special Issue provides a contribution to our understanding of salt taste and of its impact on nutrition and health. Although Na^+ is a simple chemical, the biological processes underlying its handling in our body are quite complex. Figure 1 shows a simplified scheme of sodium balance mechanisms. In the following paragraphs, I will discuss some of the new findings with relation to the processes highlighted in Figure 1: Na^+ detection, eating behavior, blood pressure regulation, and Na^+ output regulation.

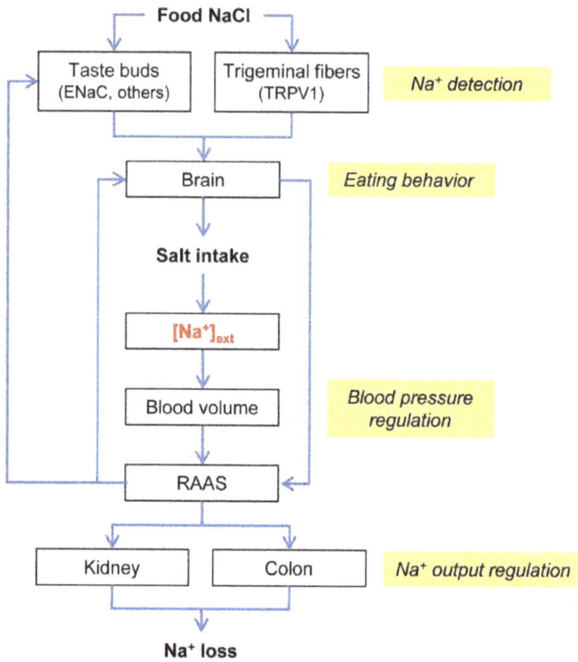

Figure 1. Sodium balance. An adequate extracellular concentration of Na^+ ($[Na^+]_{ext}$) is vital for the functioning of our cells. Diet provides the daily amount of Na^+ to balance off losses through kidney, intestine, and sweating (not shown). Food Na^+ is detected mainly by taste buds, which contained sensory cells able to sense this cation through sodium receptors (*ENaC* and others not yet identified). Taste information is fundamental for recognizing the quality and the intensity of saltiness. However, large salt concentration may also activate trigeminal nerve endings, which contain another "salt" receptor (*TRPV1*). This sensory pathway is believed to provide information on stimulus intensity in supra-threshold salt concentration range. Chemosensory signals are processed in the brain for conscious perception and for sodium appetite regulation. The outcome of central processing guides salt intake (Na^+ input). Note that oral chemosensory input provides feed-forward signals alerting central neurons on the presence of sodium-containing foods before Na^+ absorption has occurred in the gut. Once absorbed, ingested Na^+ is confined mainly to the extracellular space, and changes in its extracellular concentration affect blood volume and blood pressure, which in turn influence the renin-angiotensin-aldosterone system (*RAAS*). RAAS controls the amount of Na^+ lost by kidney and colon (Na^+ output), and also provides feedback to neural centers and taste buds to regulate sodium appetite and salt sensitivity, respectively. In turn, the brain modulates renin secretion through baroreceptor reflexes. Papers in this Special Issue touch upon some of the processes (Na^+ detection, eating behavior, blood pressure regulation, Na^+ output regulation) that are associated with the handling of food Na^+ by our body. Note that, for simplicity, other factors involved in sodium balance, such as the atrial natriuretic peptide, are not shown.

1. Na^+ Detection

The principal detectors of food Na^+ are the taste cells, specialized epithelial cells clustered in the taste buds of the oral mucosa. Na^+ activates these sensory cells by interacting with specific membrane proteins (sodium receptors) located in their apical portion, where they reach contact with the saliva. Taste cells then relay sensory information to nerve fibers, which transmit electrical impulses to the brain (Figure 1). The use of the diuretic drug, amiloride, has allowed researchers to distinguish two parallel signaling routes in most mammals: the amiloride-sensitive (AS) pathway and the amiloride-insensitive (AI) pathway. The epithelial sodium channel (ENaC), which is blocked by amiloride, works as a sodium receptor in the AS pathway. Sodium detection by the AI pathway is not affected by amiloride. The AS mechanism is highly specific for Na^+ detection, whereas the AI one is more broadly tuned to detect also other cations. However, the identity of the AI salt receptor(s) is(are) still unknown. These gustatory pathways transmit information on the stimulus quality (saltiness) and its intensity (salt concentration). Four papers provide data on the mechanisms underlying Na^+ detection by taste cells. Bigiani [1] reviews our current understanding on the possible involvement of ENaC in the initial events of Na^+ detection in humans. Although this oligomeric protein works as low-salt receptor in laboratory mammals, available data "seem to favor a role for ENaC downstream of the initial receptive events" in human salt taste. Shigemura et al. [2] provide compelling evidence that some components of the renin-angiotensin-aldosterone system (RAAS) occur in mouse taste buds. Circulating RAAS plays a key role in the regulation of sodium balance by controlling sodium excretion and by providing feedback to specific neural circuits and taste buds to regulate sodium appetite and salt sensitivity, respectively (Figure 1). The existence of components of RAAS in taste buds suggest that the activity of salt-sensitive cells could be also modulated locally in response to "random perturbations in the oral cavity during feeding and drinking", and this could affect information sent to central neurons. The authors suggest that local RAAS could be involved in "short-term feedforward regulation predicting changes in body fluid composition". Cattaneo et al. [3] found that the abundance of certain bacterial taxa on the tongue dorsum of healthy volunteers was negatively correlated with salt taste sensitivity (assessed by recognition threshold for NaCl solutions): namely, the more abundant those taxa, the lower the salt taste threshold. This result underscores the importance of the so-called perireceptor events in salt detection and possibly in defining the inter-individual differences in salt taste perception. The authors suggest that "oral microbiota composition deserves to be considered as an influencing variable when investigating perireceptor events involved in chemosensory processes". Lossow et al. [4] studied the effect of changes in dietary salt content on mRNA expression for ENaC subunits in mouse taste buds. They found that mRNA levels were not affected either by low-salt or by high-salt diet. Thus, taste function does not seem to play a major role in body adjustment to sodium imbalance, at least on the basis of the mRNA expression levels for the sodium receptor in mice.

In addition to taste cells, food Na^+ can be detected by trigeminal nerve endings, which are widely distributed throughout the lingual mucosa and also around taste buds. This sensory pathway is activated by high salt concentrations and likely provides information on salt concentration in supra-threshold range and also to avoid ingestion of hyperosmotic salt that could impair extracellular osmolality. Trigeminal nerve endings express the transient receptor potential vanilloid 1 receptor (TRPV1), an ion channel proposed to work as "salt" receptor. Rhyu et al. [5] provide interesting data on the involvement of trigeminal pathway in salt detection. These authors found that some "kokumi" peptide fractions isolated from Ganjang, a typical Korean soy sauce, were able to increase the perceived salt taste intensity in human volunteers, that is, they worked as a salt taste enhancer. Kokumi refers to tasteless compounds able to improve persistency and mouthfulness (mouth-filling sensation) as well as to enhance some of basic taste qualities, including salty taste, as demonstrated here. By recording the activity of chorda tympani (CT) taste nerve in rats, Rhyu et al. [5] found that these taste active peptides did not affect the ENaC-mediated AS pathway, but the AI one. However, the enhancing effect was likely due an interaction between trigeminal nerve endings containing TRPV1 and taste cells of the AI pathway. These results suggest "a novel relationship between trigeminal system and salt taste perception".

2. Eating Behavior

Eating behavior is a complex activity that has evolved to assure the proper ingestion of chemicals for body metabolism and homeostasis. In humans, the consumption of salty foods is driven not only by the integrated neuroendocrine mechanisms regulating the activity of central appetite neurons (Figure 1), but also by the combination of several other factors, including genetic context and non-homeostatic influences, such as learning, cultural factors, and personal habits. All these factors significantly affect the preference for salty foods, and therefore salt intake, determining individual variability.

Three papers address human eating behavior in terms of possible association between salt taste perception and salt preference or dietary habits. Cattaneo et al. [3] found a correlation between decreased salt taste sensitivity and consumption of bakery and salty baked products in healthy, normal-weight, 18-30-year old volunteers. These findings indicate that inter-individual variation in salt perception may affect habitual salt consumption. As reported above, they also found a correlation between relative abundance of certain bacterial taxa and salt taste sensitivity. Thus, oral microbiota may influence food preference. The possible association between salt taste genotype and eating behavior has been studied by Ferraris et al. [6] in a large-sized ($n = 536$) and well-characterized elderly cohort in Australia. They evaluated the association between the occurrence of single nucleotide polymorphisms (SNPs) for the TRPV1-encoding gene and individual salt intake. TRPV1 is a salt receptor expressed in trigeminal nerve endings, which may contribute to the perception of NaCl, especially when salt concentration is high. Their findings indicate that there is no association between TRPV1 SNP and salt intake in the analyzed elderly cohort, suggesting that, at least for people aged 65 years or older, the TRPV1 genotype is not crucial in defining salt consumption. Veček et al. [7] performed a cross-sectional study on general population of Dalmatia, Croatia ($n = 2798$ subjects) to determine possible association between salt taste perception, Mediterranean diet, and Metabolic Syndrome (MetS). They found that there were no differences in the overall Mediterranean diet compliance between subjects with different salt taste threshold. However, they found that "subjects with higher salt threshold added salt to their food more frequently compared to subjects with both lower and intermediate threshold". This means that individuals with higher salt sensitivity (lower taste threshold) could have a reduced salt intake. Interestingly, they found that these subjects also showed lower prevalence of MetS.

3. Blood Pressure Regulation

Nervous and endocrine mechanisms assure a proper level of hydraulic pressure in the large arteries to sustain an adequate blood perfusion throughout the body. Salt intake may affect blood pressure. Ingested Na^+ is confined into the extracellular compartment. Here, through water retention, Na^+ sets the overall volume, including the blood volume. In turn, this affects blood pressure (Figure 1). It is therefore clear that one of the consequences of an increased salt intake is likely an increase in blood pressure. Several studies underscore the role of excessive salt consumption in the development of hypertension, which represents a component of MetS. The association between salt intake and blood pressure is addressed in this Special Issue. Ferraris et al. [6] found (see above) that SNP for the gene encoding the salt receptor, TRPV1, was not associated with a variation in salt intake in an elderly cohort. Consistently, neither systolic nor diastolic blood pressure varied by genotype. The authors acknowledge that "as the cohort was 65 years and older, the results are not necessarily generalizable to the wider adult or youth population" since aging may affect genetic expression. In their cross-sectional study on the general population of Dalmatia (Croatia), Veček et al. [7] found that "subjects with higher salt taste threshold were on average older than those with lower threshold". This means that the ability to recognize salty stimuli blunts with age. In addition, they found that "age was also negatively correlated with salt taste intensity perception", evaluated with supra-threshold testing. As a whole, these results suggest that aging affects salt taste performance. Veček et al. [7] also found that high blood pressure, a MetS component, was more common among subjects with higher salt recognition threshold, that is, those with lower salt sensitivity. Mun et al. [8] studied the effect of Doenjang, a traditional Korean seasoning with a high salt content obtained by soybean fermentation, on the blood pressure in rats, which were fed a high-salt diet with or without Doenjang. Interestingly, blood pressure was significantly lower in the first group, although the administered salt content was similar in the two groups. In addition, RAAS was also affected: renin and aldosterone levels were decreased in mice fed with Doenjang. Likely, other chemicals occurring in Doenjang as well as its microbial community offset the effect of salt intake on blood pressure. Thus, their results suggest that "eating traditional salty fermented food is not a direct cause of hypertension, and the intake of Doenjang in normal healthy animals improved blood pressure". In line with the goal of reducing salt consumption without affecting the palatability of foodstuffs, Rhyu et al. [5] identified kokumi active peptides that could be used as salt taste enhancers. Clearly, this would be beneficial for controlling blood pressure, because less salt would be necessary to have the same saltiness perception when these peptides are present in foods.

4. Na^+ Output Regulation

Renal excretion and losses through the gastrointestinal tract represent the output of the homeostatic system controlling the extracellular concentration of Na^+ ($[Na^+]_{ext}$). Variations in salt intake lead to changes in $[Na^+]_{ext}$, which in turn affect, through expansion/reduction of extracellular volume to correct osmolality, blood volume and RAAS (Figure 1). RAAS targets the renal nephron and the colon, where aldosterone affects sodium reabsorption through ENaC to match changes in sodium intake. Lossow et al. [4] found that changes in dietary salt content in mice affected ENaC mRNA expression levels in kidney and distal colon. As reported above, they also analyzed ENaC expression in taste buds, since in this mammal ENaC works as sodium taste receptor. They data clearly indicate that "colon and kidney seem to be of greater importance to compensate imbalanced sodium homeostasis than gustatory tissue based on the monitored ENaC expression levels". These results underscore the importance of an adequate regulation of sodium output to compensate changes in $[Na^+]_{ext}$ due to variations in salt intake. In this regard, it is worth noting that salt taste works as a feedforward mechanism that can reduce sodium appetite to prevent overconsumption. On the contrary, Na^+ output is the expression of a homeostatic regulation, which uses compensatory feedback mechanisms to stabilize $[Na^+]_{ext}$ (Figure 1).

5. Final Note

I would like to thank all the authors in this Special Issue for providing their new research data on salt taste mechanisms and on the role of salt taste in nutrition and health. I am sure their valuable contributions will be appreciated by the readership of *Nutrients*, as well as by the scientific community.

Funding: This research received no external funding.

Conflicts of Interest: The author declares no conflict of interest.

References

1. Bigiani, A. Does ENaC work as sodium taste receptor in humans? *Nutrients* **2020**, *12*, 1195. [CrossRef] [PubMed]
2. Shigemura, N.; Takai, S.; Hirose, F.; Yoshida, R.; Sanematsu, K.; Ninomiya, Y. Expression of renin-angiotensin system components in the taste organ of mice. *Nutrients* **2019**, *11*, 2251. [CrossRef] [PubMed]
3. Cattaneo, C.; Riso, P.; Laureati, M.; Gargari, G.; Pagliarini, E. Exploring associations between interindividual differences in taste perception, oral microbiota composition, and reported food intake. *Nutrients* **2019**, *11*, 1167. [CrossRef] [PubMed]
4. Lossow, K.; Meyerhof, W.; Behrens, M. Sodium imbalance in mice results primarily in compensatory gene regulatory responses in kidney and colon, but not in taste tissue. *Nutrients* **2020**, *12*, 995. [CrossRef] [PubMed]
5. Rhyu, M.-R.; Song, A.-Y.; Kim, E.-Y.; Son, H.-J.; Kim, Y.; Mummalaneni, S.; Qian, J.; Grider, J.R.; Lyall, V. Kokumi taste active peptides modulate salt and umami taste. *Nutrients* **2020**, *12*, 1198. [CrossRef] [PubMed]
6. Ferraris, C.; Turner, A.; Kaur, K.; Piper, J.; Veysey, M.; Lucock, M.; Beckett, E.L. Salt taste genotype, dietary habits and biomarkers of health: No associations in an elderly cohort. *Nutrients* **2020**, *12*, 1056. [CrossRef] [PubMed]
7. Veček, N.N.; Mucalo, L.; Dragun, R.; Miličević, T.; Pribisalić, A.; Patarčić, I.; Hayward, C.; Polašek, O.; Kolčić, I. The association between salt taste perception, mediterranean diet and metabolic syndrome: A cross-sectional study. *Nutrients* **2020**, *12*, 1164. [CrossRef] [PubMed]
8. Mun, E.G.; Park, J.E.; Cha, Y.S. Effects of Doenjang, a traditional Korean soybean paste, with high-salt diet on blood pressure in Sprague-Dawley rats. *Nutrients* **2019**, *11*, 2745. [CrossRef] [PubMed]

nutrients

MDPI

Review

Does ENaC Work as Sodium Taste Receptor in Humans?

Albertino Bigiani

Dipartimento di Scienze Biomediche, Metaboliche e Neuroscienze, Università di Modena e Reggio Emilia, 41125 Modena, Italy; albertino.bigiani@unimore.it; Tel.: +39-059-205-5349

Received: 29 March 2020; Accepted: 21 April 2020; Published: 24 April 2020

check for updates

Abstract: Taste reception is fundamental for the proper selection of food and beverages. Among the several chemicals recognized by the human taste system, sodium ions (Na^+) are of particular relevance. Na^+ represents the main extracellular cation and is a key factor in many physiological processes. Na^+ elicits a specific sensation, called salty taste, and low-medium concentrations of table salt (NaCl, the common sodium-containing chemical we use to season foods) are perceived as pleasant and appetitive. How we detect this cation in foodstuffs is scarcely understood. In animal models, such as the mouse and the rat, the epithelial sodium channel (ENaC) has been proposed as a key protein for recognizing Na^+ and for mediating preference responses to low-medium salt concentrations. Here, I will review our current understanding regarding the possible involvement of ENaC in the detection of food Na^+ by the human taste system.

Keywords: sodium taste; sodium receptor; salt taste; amiloride; taste transduction

1. Introduction

The sodium ion (Na^+) is an essential mineral for our body because it regulates the osmolality of the extracellular fluid and plays a key role in many physiological processes, from the generation of nerve impulses to renal function. Na^+ is lost continuously through urine, feces, and sweat. Thus, to maintain proper bodily balance, we need to replace losses by the ingestion of food containing this cation. Our ability to detect Na^+ in foodstuffs relies on the taste system: Na^+ elicits a specific sensation called salty taste [1,2] that guides the intake of this important mineral [3,4]. Table salt (NaCl) represents a prototypic chemical substance evoking salty taste. It is well established that Na^+ is responsible for the perceived saltiness and for the pleasantness of low to medium concentrations of table salt [4,5].

The early events in taste reception typically include the interaction of the chemical stimulus with a membrane receptor in taste cells, specialized epithelial cells clustered in sensory end-organs called taste buds [6]. Most of studies on sodium taste reception have been performed on laboratory rodents (mouse and rat). In these mammals, the epithelial sodium channel (ENaC) works as low-salt receptor, mediating acceptance responses to low-medium salt concentrations and driving salt consumption [5]. The obvious question is then: does ENaC play any role in sodium detection in humans? In this review, I will discuss current information supporting or arguing against the possible involvement of this ion channel in human salt taste. I will evaluate whether data from human studies fit the model for the detection of Na^+ based on ENaC as sodium receptor localized at the apical membrane of taste cells; that is, where these cells contact the saliva in the so-called taste pore region of taste buds [6,7].

2. Psychophysics

In laboratory rodents, a pharmacological feature of taste ENaC is its sensitivity to amiloride, a diuretic drug, which selectively blocks the channel at submicromolar concentrations [8]. Since ENaC is inhibited by amiloride, application of this drug during stimulation with NaCl blunts the taste response and the attractiveness of low sodium concentrations [9,10]. Following the same methodological approach, the involvement of the ENaC pathway in human taste reception has been investigated over 15 years by sensory evaluation of the amiloride effect on salt taste. Unexpectedly, findings have been very controversial. Some authors found that indeed the presence of amiloride reduced the perceived saltiness of the NaCl solutions, although to varying degrees [11–15]. On the contrary, other investigators found that amiloride had negligible or no effect on salt taste [16–21]. The discrepancy among studies might be due to differences in the experimental design. For example, perception arising from stimulation of the anterior tongue seems to be affected by amiloride [11–13], whereas perception from whole-mouth stimulation is not [16]. This raises the possibility that in the whole-mouth protocol, additional sensory inputs from the oral cavity may "obscure" the information conveyed by amiloride-sensitive pathway at the level of central processing [22]. However, other studies in which stimuli were delivered only to the anterior dorsal surface of the tongue failed to find a significant effect of amiloride on saltiness perception [17,20,21]. Of note, amiloride strongly reduces the sour side taste of salt solutions when subjects can use not only one response category (saltiness), but all taste qualities (saltiness, sweetness, sourness, bitterness) to describe their perception [17,19].

Further factors affecting the outcome of the sensory assays might include the impact of amiloride bitterness in establishing the taste quality of salt solutions as well as the amiloride concentration tested. It is worth noting that some studies showing the amiloride effect used high drug concentrations (500 μM and 1 mM) [12–14], whereas studies claiming a negligible effect of amiloride on salt perception adopted smaller concentrations of the drug (10–100 μM) [17,18,20,21]. The specificity of amiloride for ENaC is true only for low drug concentration (half-maximal inhibition in submicromolar range; [8]). At higher concentrations, amiloride also affects other cellular proteins, including ion channels, transporters, and receptors [3,23]. A few examples of the molecular targets other than ENaC are shown in Table 1. It is worth noting the same proteins are also found in the taste cells of laboratory rodents (Table 1, rightmost column). As nicely pointed out by Lindemann [3] "if concentrations above 10 μM are needed for half-maximal inhibition, effects of amiloride other than blockage of Na^+ channels need to be considered". Thus, it is possible that human testing may have yielded contrasting results due to nonspecific effects of amiloride. However, it is also possible that ENaC in humans displays a lower sensitivity to amiloride than typical ENaC, as indicated by molecular biology studies (see below).

Table 1. Molecular targets, other than ENaC, for the inhibitory effect of amiloride expressed by the inhibition constant (K_i = amiloride concentration producing 50% inhibition). Data refer to nontaste tissues. The occurrence of these molecular targets in rodents taste cells is also referenced.

Molecular Target	K_i (μM)	Cell/Tissue	Occurrence in Rodent Taste Cells
T-type calcium channel	30	Mouse neuroblastoma and chick DRG [1] neurons [24]	[25,26]
Na^+/H^+ exchanger	30	Rabbit renal microvillous membrane [27]	[28–30]
Muscarinic receptors	40–80	Rabbit pancreatic acini [31]	[32,33]

[1] Dorsal Root Ganglion.

3. Electrophysiology

In laboratory animals, application of NaCl solution to the tongue mucosa produces a transepithelial current due to the movement of ions across the epithelium [9,34,35]. This current is believed to be sustained mainly by Na^+ entering taste cells via the apical ENaC since amiloride strongly reduces it. Obviously, during NaCl stimulation, a voltage drop between mucosal and serosal side of the tongue develops, and this voltage drop can be measured with adequate instrumentation [9,34,35]. This approach has been applied to human volunteers to evaluate the contribution of the amiloride-sensitive pathway to the perceived saltiness. Electrophysiological recordings of lingual surface potential (LSP) in response to focal NaCl stimulation have provided evidence that, in some individuals, amiloride (100 μM) was able to reduce the voltage drop caused by NaCl application [36]. However, the effect was highly variable among individuals, ranging from 0% to 42% inhibition. Further studies demonstrated a positive correlation between LSP and the perceived intensity of saltiness [37]. It was also found that in expert salt tasters, amiloride (10 μM) impaired the ability to distinguish between two different salt concentrations (100 and 300 mM NaCl) [37]. Again, not all subjects exhibiting a LSP during NaCl stimulation were sensitive to amiloride, underscoring the high variability of the amiloride effect across individuals observed previously [36]. Nonetheless, these electrophysiological studies seem to confirm the psychophysical findings suggesting suppression of perceived saltiness by amiloride in some individuals when a small area of the tongue surface is stimulated [11–13].

4. Molecular Biology and Immunohistochemistry

In laboratory rodents, ENaC is an oligomeric protein made of three nonidentical subunits, named α-, β-, and γ-ENaC [38,39]. Expression cloning studies with *Xenopus* oocytes have clearly indicated that the α-subunit is required to induce channel activity, whereas the presence of the β- and γ-subunit allows maximal expression of sodium current [40]. Although it is still unknown how these subunits assemble to form ENaC in taste cells [41], all of them have been detected in taste tissues from laboratory animals by using molecular and immunohistochemical techniques [42–46]. It is important to underscore that there is also direct evidence that amiloride-sensitive taste cells do have ENaC subunits, whereas amiloride-insensitive cells do not [46]. By applying the same techniques on tissue samples from human subjects, it has been possible to establish that α-, β-, and γ-ENaCs occur in human taste papillae [47,48]. In man, however, taste tissues also express an additional δ-subunit, which is missing in rodents [48,49]. The δ-subunit is analog to the α-subunit in that its presence is necessary to form a Na^+-permeable channel [50]. It is then possible that in human taste tissues, ENaC may include either an α- or δ-subunit. This subunit change may have an impact on the amiloride sensitivity, since replacement of α-subunit by a δ-subunit makes the channel 50-fold less sensitive to amiloride [50–52]. If this is the case, then the negligible effect of amiloride observed in some psychophysical studies (e.g., [17,18]) could be attributed to variations in the molecular composition of the ENaC protein [53]. It is worth noting that the expression level of ENaC subunits may vary significantly among subjects and that ENaC mRNAs are also detected in nontaste epithelium [48].

The detection of ENaC subunits in human taste tissues does not necessarily imply that this channel is involved in the initial events of sodium detection. According to the model of sodium taste detection proposed for rodents, ENaC should be found at the apical membrane of taste cells to work as a sodium receptor [3,8]. Immunohistochemical localization of ENaC subunits in human lingual epithelium has revealed that δ-ENaC is exclusively restricted to the taste pore region in both fungiform and circumvallate taste buds [48]. However, it has not been possible to establish whether this subunit localized to the apical membrane of taste cells or to tight junctions surrounding the apical ends of these cells. Unexpectedly, the other ENaC-subunits were found in the basolateral compartment of taste cells, which is involved in later stages of the sensory transduction and in intercellular communications with nerve endings and adjacent cells [5]. This raises the possibility that ENaC may serve other functions in taste buds. Recent findings indicate that ENaC plays a key role in the regulation of adult neurogenesis [54]. It is then tempting to speculate that ENaC might be involved in taste cell development since these cells continuously turnover [55,56]. Of note, another study found that δ-ENaC immunoreactivity was distributed over both the apical and basolateral ends of fungiform taste cells [49]. Although it is not clear how to reconcile these conflicting results on labeling pattern, both findings support the notion that δ-ENaC is expressed in human taste buds.

Although the model of sodium detection involving apically located ENaC explains several experimental observations, ENaC-subunits localized to the basolateral membrane could also mediate sodium reception by sensing Na^+ leaked through tight junctions around taste cells [5]. Thus, this paracellular pathway could be responsible for activation of taste cells even in the absence of apical ENaC. Since the basolateral compartment of taste cells is exposed to an extracellular solution containing about 150 mM Na^+, a significant diffusion of this cation through tight junctions could occur only if Na^+ concentration in the mucosal surface is much higher than 150 mM. It is then possible that basolateral ENaCs may be relevant for sodium detection when salt concentration in the stimulating solution exceeds plasma tonicity. The basolateral localization of ENaC subunits may be a further factor in determining the variable amiloride sensitivity observed in human studies (see above).

5. Genetics

Single nucleotide polymorphisms (SNPs) in the gene coding for the ENaC β-subunit (*SCNN1B*) are somehow associated with changes in suprathreshold taste sensitivity for NaCl solutions, but not with salt taste threshold [57]. The β-subunit does not play a role in pore formation of the channel protein, but it is assumed to modulate channel activity and to be important for channel trafficking to the cell membrane [40,58,59]. Thus, these findings indicate that variations in the β-ENaC genes may contribute to differences in salt taste perception among individuals through a possible effect on the expression of ENaC in the taste cell membrane. Interestingly, these data imply that ENaC may be involved in the recognition of NaCl at concentrations that would have an impact on the actual consumption of dietary salt, that is, at suprathreshold concentrations. As pointed out by Contreras [60], in general, people do not add salt to food in order to be able to just detect it, but do so to a preferred suprathreshold level. It is noteworthy that amiloride tends to reduce suprathreshold intensities of perceived NaCl in adult volunteers [15]. However, a significant difference in taste intensity ratings between individuals with SNPs was found only for large concentrations of NaCl, such as 1 M [57]. Aqueous solutions of salt above ~150 mM are not preferred by humans [61]. In animals, salt levels exceeding tonicity of blood plasma are normally not accepted [62]. High salt concentrations activate other sensory pathways in addition to the ENaC-mediated one, including an amiloride-insensitive taste pathway and trigeminal nerve endings [5,22]. These components of salt reception mediate aversion responses and work as warning mechanisms to avoid the ingestion of hyperosmotic salt solutions [5,10,63]. Thus, the findings on SNPs of the ENaC β-subunit seemingly do not fit the model involving ENaC as low-salt receptor.

6. Salt Taste Enhancers

Chemicals able to increase the sensation evoked by NaCl without being salty themselves, the so-called "salt taste enhancers", have attracted the attention of researchers for many years [64]. The reason is that these substances may be used to reduce salt content in processed foods to prevent excessive sodium intake, which is linked to the development of hypertension and subsequent pathologies [65,66]. Research on the mechanisms underlying the action of salt taste enhancers has provided some clues on the peripheral events leading to salt taste perception in humans.

Studies on human αβγ- or δβγ-ENaC functionally expressed in *Xenopus* oocytes have shown that sodium current through ENaCs is activated by salt-taste-modulating substances, such as L-arginine (Arg) [48]. By monitoring the changes of intracellular calcium levels in cultured human fungiform taste (HBO) cells, Xu et al. [67] found recently that some arginyl dipeptides, which proved to work as potent salty taste enhancers (up to 20% increase in perceived saltiness [68]), induced a significant increase in the number of cultured cells responding to NaCl. They also found that the effect required the presence of either α-ENaC or δ-ENaC. These results clearly indicate that salt taste enhancers target the human sodium receptor ENaC in both the αβγ or δβγ form. Unexpectedly, Arg was unable to stimulate cultured human taste cells [67], although it enhances the perceived saltiness in human sensory evaluations [48,68].

There are some aspects of the above studies that require keen consideration. In particular, it is remarkable how the effect of Arg may be affected by the cell system used to express human αβγ- or δβγ-ENaC. In *Xenopus* oocytes, Arg increases the sodium current through ENaCs [48], whereas in HBO cells, it is ineffective in changing intracellular calcium levels [67]. It is possible that this discrepancy may derive from the different experimental and methodological approach adopted. However, the finding that Arg potentiates ion currents through ENaC is, by itself, quite surprising. Both Arg and amiloride bear a guanidinium group (Figure 1), which occurs as a cation in physiological conditions (pH 7.4). It is believed that the positive charge-bearing guanidinium group of amiloride penetrates part of the ENaC channel pore, causing channel blockage, whereas the pyrazine group interacts with the outer mouth of the channel [69]. The chemical similarity may suggest that Arg, like amiloride, could affect ENaC directly from the extracellular space. Indeed, Ogawa et al. [70] suggested that "the guanidinium group of Arg may interact with sodium channels in taste bud membranes". Yet, Arg enhances the current through ENaC, whereas amiloride reduces it.

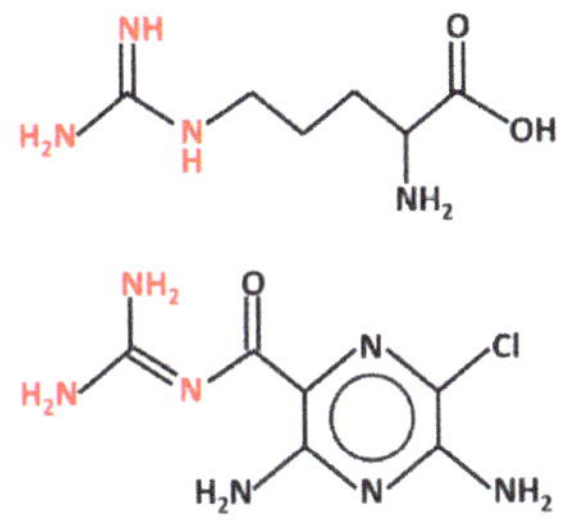

Figure 1. Structure of L-arginine (top) and amiloride (bottom). Both chemicals bear a guanidinium group (red), which is protonated in physiological conditions. This group is believed to interact with the ENaC channel pore from the extracellular space.

7. Salivary Proteins

Proteins represent an important component of the saliva [71], the medium carrying Na^+ to the apical, chemosensitive tips of taste cells. Recent studies on human subjects have found a correlation between salivary serine-type endoprotease activity and sensitivity to NaCl [72]. Since serine proteases increase the activity of ENaC through proteolytic cleavage [73,74], the authors have proposed that endoproteases of the saliva might affect salt taste sensitivity by modifying ENaC functioning. Stolle et al. [75] have identified a tetrapeptide that is likely released from salivary proteins by serine-type endoprotease activity and that is able to enhance salt taste perception. This means that in the saliva of salt sensitive subjects, an endogenous salt enhancer might be produced by enzymatic cleavage. It was also found that the abundance of two salivary proteins, lipocalin-1 and lysozyme C, could be related to individuals' low sensitivity to NaCl. The authors have put forward the hypothesis that electrostatic interaction of these proteins with ENaC in taste cells may reduce the accessibility of sodium ions to ENaC [75]. These findings are clearly fascinating, but do not prove that ENaC actually functions as a sodium receptor in humans.

8. Discussion

The possible involvement of the ion channel, ENaC, in human taste reception has been investigated with different approaches. To date, however, it is not possible to provide a definitive answer as to the role of the ENaC pathway in producing salty sensations due to inconsistent findings. Perhaps the more conflicting results are from psychophysical studies involving the use of amiloride, a pharmacological probe for ENaC, to challenge saltiness perception. There are several issues regarding the adopted methodology that might be responsible for the observed discrepancies. For instance, the amiloride concentration used in sensory tests is not always adequate to avoid side effects on other ion channels and transporters. Thus, the apparent effect of amiloride in some studies (e.g., [12–14]) might be somehow misleading. Nonetheless, electrophysiological studies suggest that the application of salt solution on the human tongue induces a voltage drop across the mucosa that is similar to the one observed in laboratory animals. However, ENaC mRNAs are found also in nontaste epithelium [44,48,76], raising the possibility that the ion current crossing the mucosa could also be due to Na^+ diffusion through epithelial cells.

According to the model proposed for rodents, ENaC should be found at the apical membrane of taste cells, which stick out into the taste pore bathed by saliva containing taste stimuli [3,8]. Available data suggest that only the δ-subunit localizes to the taste pore region in human taste buds, whereas other ENaC subunits seem to be segregated in the basolateral compartment, beneath the apical tips of taste cells. It is not yet known whether all the subunits are required to form a functional sodium receptor [41]. Differential expression of ENaC-subunits has been described in transporting epithelia [77], suggesting that endogenous channel in vivo may require only one or two subunits to work properly. δ-ENaC expressed alone in *Xenopus* oocytes is able to mediate a membrane current [50]. Thus, the occurrence of the δ-subunit in the taste pore region of human taste buds seems to suggest that it may function as a sodium receptor. Unfortunately, the microscopic analysis has not allowed establishing with confidence whether this subunit lies in the apical end of taste cells or in the tight junctions connecting adjacent taste cells just below the taste pore [48]. Indirect evidence supporting a role for ENaC in human salt taste has been provided by genetic studies [57] and by in vitro assays on cells expressing the human αβγ- or δβγ-ENaC [48,67], although there is some inconsistency among these data. Recent analysis of the correlation between the salivary proteome and the salt sensitivity in human volunteers are seemingly consistent with a role of ENaC [72,75].

In conclusion, the available data are suggestive of possible involvement of ENaC in human sodium detection, although it is not clear whether this occurs at the beginning of the reception process (interaction between sodium receptor and Na^+ at the apical membrane of taste cells) or later on, after Na^+ has been detected. The lack of the amiloride effect in some psychophysical studies [16–18,20,21] and the presence of α-, β-, and γ-subunit only in the basolateral portion of taste buds [48] seem to favor a role for ENaC downstream of the initial receptive events. Consistent with this hypothesis is the finding that SNPs in the gene coding for the ENaC β-subunit affect suprathreshold sensitivity to salt solutions, that is, at concentration levels above the detection/recognition threshold [57]. Clearly, further research is required to obtain a coherent and thorough comprehension of the early events of sodium detection in human taste cells. This information represents the premise for understanding interindividual variability in the function of sodium taste receptors and its potential implications for eating behavior.

Funding: This research received no external funding.

Acknowledgments: This work was supported in part by Università di Modena e Reggio Emilia (FAR Dipartimentale 2018, 2019).

Conflicts of Interest: The author declares no conflict of interest.

References

1. Schiffman, S.S.; Erickson, R.P. A psychophysical model for gustatory quality. *Physiol. Behav.* **1971**, *7*, 617–633. [CrossRef]
2. Mattes, R.D. The taste for salt in humans. *Am. J. Clin. Nutr.* **1997**, *65*, 692S–697S. [CrossRef]
3. Lindemann, B. Taste reception. *Physiol. Rev.* **1996**, *76*, 718–766. [CrossRef]
4. McCaughey, S.A.; Scott, T.R. The taste of sodium. *Neurosci. Biobehav. Rev.* **1998**, *22*, 663–676. [CrossRef]
5. Bigiani, A. Salt taste. In *The Senses. Olfaction and Taste*, 2nd ed.; Meyerhof, W., Ed.; Elsevier: Amsterdam, The Netherlands, 2020; Volume III, in press. [CrossRef]
6. Breslin, P.A.; Huang, L. Human taste: Peripheral anatomy, taste transduction, and coding. *Adv. Otorhinolaryngol.* **2006**, *63*, 152–190. [CrossRef]
7. Witt, M. Anatomy and development of the human taste system. In *Handbook of Clinical Neurology. Smell and Taste*; Doty, R.L., Ed.; Elsevier: Amsterdam, The Netherlands, 2019; Volume 164, pp. 147–171.
8. Bigiani, A. Electrophysiology of sodium receptors in taste cells. *J. Biomed. Sci. Eng.* **2016**, *9*, 367–383. [CrossRef]
9. Heck, G.L.; Mierson, S.; DeSimone, J.A. Salt taste transduction occurs through an amiloride-sensitivesodium transport pathway. *Science* **1984**, *223*, 403–405. [CrossRef]
10. Chandrashekar, J.; Kuhn, C.; Oka, Y.; Yarmolinsky, D.A.; Hummler, E.; Ryba, N.J.P.; Zuker, C.S. The cells and peripheral representation of sodium taste in mice. *Nature* **2010**, *464*, 297–301. [CrossRef]

11. Schiffman, S.S.; Lockhead, E.; Maes, F.W. Amiloride redcuces the taste intensity of Na^+ and Li^+ salts and sweeteners. *Proc. Natl. Acad. Sci. USA* **1983**, *80*, 6136–6140. [CrossRef]
12. Tennissen, A.M. Amiloride reduces intensity responses of human fungiform papillae. *Physiol. Behav.* **1992**, *51*, 1061–1068. [CrossRef]
13. McCutcheon, N.B. Human psychophysical studies of saltiness suppression by amiloride. *Physiol. Behav.* **1992**, *51*, 1069–1074. [CrossRef]
14. Tennissen, A.M.; McCutcheon, N.B. Anterior tongue stimulation with amiloride suppress NaCl saltiness, but not citric acid sourness in humans. *Chem. Senses* **1996**, *21*, 113–120. [CrossRef]
15. Anand, K.K.; Zuniga, J.R. Effect of amiloride on suprathreshold NaCl, LiCl, and KCl salt taste in humans. *Physiol. Behav.* **1997**, *62*, 925–929. [CrossRef]
16. Desor, J.A.; Finn, J. Effects of amiloride on salt taste in humans. *Chem. Senses* **1989**, *14*, 793–803. [CrossRef]
17. Ossebaard, C.A.; Smith, D.V. Effect of amiloride on the taste of NaCl, Na-gluconate and KCl in humans: Implications for Na^+ receptor mechanisms. *Chem. Senses* **1995**, *20*, 37–46. [CrossRef]
18. Smith, D.V.; Ossebaard, C.A. Amiloride suppression of the taste intensity of sodium chloride: Evidence from direct magnitude scaling. *Physiol. Behav.* **1995**, *4*, 773–777. [CrossRef]
19. Ossebaard, C.A.; Smith, D.V. Amiloride suppresses the sourness of NaCl and LiCl. *Physiol. Behav.* **1996**, *60*, 1317–1322. [CrossRef]
20. Ossebaard, C.A.; Polet, I.A.; Smith, D.V. Amiloride effects on taste quality: Comparison of single and multiple response category procedures. *Chem. Senses* **1997**, *22*, 267–275. [CrossRef]
21. Halpern, B.P.; Darlington, R.B. Effects of amiloride on gustatory quality descriptions and temporal patterns produced by NaCl. *Chem. Senses* **1998**, *23*, 501–511. [CrossRef]
22. Simon, S.A.; de Araujo, I.E.; Stapleton, J.R.; Nicolelis, M.A. Multisensory processing of gustatory stimuli. *Chemosens. Percept.* **2008**, *1*, 95–102. [CrossRef]
23. Kleyman, T.R.; Cragoe, E.J., Jr. Amiloride and its analogs as tools in the study of ion transport. *J. Membr. Biol.* **1988**, *105*, 1–21. [CrossRef] [PubMed]
24. Tang, C.-M.; Presser, F.; Morad, M. Amiloride selectively blocks the low threshold (T) calcium channel. *Science* **1988**, *240*, 213–215. [CrossRef] [PubMed]
25. Béhé, P.; DeSimone, J.A.; Avenet, P.; Lindemann, B. Membrane currents in taste cells of the rat fungiform papilla. *J. Gen. Physiol.* **1990**, *96*, 1061–1084. [CrossRef] [PubMed]
26. Bigiani, A.; Cuoghi, V. Localization of amiloride-sensitive sodium current and voltage-gated calcium currents in rat fungiform taste cells. *J. Neurophysiol.* **2007**, *98*, 2483–2487. [CrossRef]
27. Kinsella, J.L.; Aronson, P.S. Amiloride inhibition of the Na^+-H^+ exchanger in renal microvillus membrane vesicles. *Am. J. Physiol.* **1981**, *241*, F374–F379. [CrossRef]
28. Lundy, R.F., Jr.; Pittman, D.W.; Contreras, R.J. Role for epithelial Na^+ channels and putative Na^+/H^+ exchangers in salt taste transduction in rats. *Am. J. Physiol.* **1997**, *273*, R1923–R1931. [CrossRef]
29. Vinnikova, A.K.; Alam, R.I.; Malik, S.A.; Ereso, G.L.; Feldman, G.M.; McCarty, J.M.; Knepper, M.A.; Heck, G.L.; DeSimone, J.A.; Lyall, V. Na^+-H^+ exchange activity in taste receptor cells. *J. Neurophysiol.* **2004**, *91*, 1297–1313. [CrossRef] [PubMed]
30. Lyall, V.; Alam, R.I.; Malik, S.A.; Phan, T.H.; Vinnikova, A.K.; Heck, G.L.; DeSimone, J.A. Basolateral Na^+-H^+ exchanger-1 in rat taste receptor cells is involved in neural adaptation to acidic stimuli. *J. Physiol.* **2004**, *556*, 159–173. [CrossRef]
31. Kuijpers, G.A.; De Pont, J.J.; Van Nooy, I.G.; Fleuren-Jakobs, A.M.; Bonting, S.L.; Rodrigues de Miranda, J.F. Amiloride is a cholinergic antagonist in the rabbit pancreas. *Biochim. Biophys. Acta* **1984**, *804*, 237–244. [CrossRef]
32. Ogura, T. Acetylcholine increases intracellular Ca^{2+} in taste cells via activation of muscarinic receptors. *J. Neurophysiol.* **2002**, *87*, 2643–2649. [CrossRef]
33. Eguchi, K.; Ohtubo, Y.; Yoshii, K. Functional expression of M3, a muscarinic acetylcholine receptor subtype, in taste bud cells of mouse fungiform papillae. *Chem. Senses* **2008**, *33*, 47–55. [CrossRef] [PubMed]
34. DeSimone, J.A.; Heck, G.L.; DeSimone, S.K. Active ion transport in dog tissue: A possible role in taste. *Science* **1981**, *214*, 1039–1041. [CrossRef] [PubMed]
35. DeSimone, J.A.; Heck, G.L.; Mierson, S.; DeSimone, S.K. The active ion transport properties of canine lingual epithelia in vitro. Implications forgustatory transduction. *J. Gen. Physiol.* **1984**, *83*, 633–656. [CrossRef]

36. Feldman, G.M.; Mogyorósi, A.; Heck, G.L.; DeSimone, J.A.; Santos, C.R.; Clary, R.A.; Lyall, V. Salt-evoked lingual surface potential in humans. *J. Neurophysiol.* **2003**, *90*, 2060–2064. [CrossRef] [PubMed]
37. Feldman, G.M.; Heck, G.L.; Smith, N.L. Human salt taste and the lingual surface potential correlate. *Chem. Senses* **2009**, *34*, 373–382. [CrossRef] [PubMed]
38. Benos, D.J.; Stanton, B.A. Functional domains within the degenerin/epithelial sodium channel (Deg/ENaC) superfamily of ion channels. *J. Physiol.* **1999**, *520*, 631–644. [CrossRef] [PubMed]
39. Kellenberger, S.; Schild, L. International Union of Basic and Clinical Pharmacology. XCI. structure, function, and pharmacology of acid-sensing ion channels and the epithelial Na^+ channel. *Pharmacol. Rev.* **2015**, *67*, 1–35. [CrossRef]
40. Canessa, C.M.; Schild, L.; Buell, G.; Thorens, B.; Gautschi, I.; Horisberger, J.D.; Rossier, B.C. Amiloride-sensitive epithelial Na^+ channel is made of three homologous subunits. *Nature* **1994**, *367*, 463–467. [CrossRef]
41. Lossow, K.; Hermans-Borgmeyer, I.; Meyerhof, W.; Behrens, M. Segregated expression of ENaC subunits in taste cells. *Chem. Senses* **2020**, in press. [CrossRef]
42. Simon, S.A.; Holland, V.F.; Benos, D.J.; Zampighi, G.A. Transcellular and paracellular pathways in lingual epithelia and their influence in taste transduction. *Microsc. Res. Tech.* **1993**, *26*, 196–208. [CrossRef]
43. Kretz, O.; Barbry, P.; Bock, R.; Lindemann, B. Differential expression of RNA and protein of the three pore-forming subunits of the amiloride-sensitive epithelial sodium channel in taste buds of the rat. *J. Histochem. Cytochem.* **1999**, *47*, 51–64. [CrossRef] [PubMed]
44. Lin, W.; Finger, T.E.; Rossier, B.C.; Kinnamon, S.C. Epithelial Na^+ channel subunits in rat taste cells: Localization and regulation by aldosterone. *J. Comp. Neurol.* **1999**, *405*, 406–420. [CrossRef]
45. Shigemura, N.; Islam, A.A.; Sadamitsu, C.; Yoshida, R.; Yasumatsu, K.; Ninomiya, Y. Expression of amiloride-sensitive epithelial sodium channels in mouse taste cells after chorda tympani nerve crush. *Chem. Senses* **2005**, *30*, 531–538. [CrossRef] [PubMed]
46. Yoshida, R.; Horio, N.; Murata, Y.; Yasumatsu, K.; Shigemura, N.; Ninomiya, Y. NaCl responsive taste cells in the mouse fungiform taste buds. *Neuroscience* **2009**, *159*, 795–803. [CrossRef]
47. Rossier, O.; Cao, J.; Huque, T.; Spielman, A.I.; Feldman, R.S.; Medrano, J.F.; Brand, J.G.; le Coutre, J. Analysis of a human fungiform papillae cDNA library and identification of taste-related genes. *Chem. Senses* **2004**, *29*, 13–23. [CrossRef]
48. Stähler, F.; Riedel, K.; Demgensky, S.; Neumann, K.; Dunkel, A.; Täubert, A.; Raab, B.; Behrens, M.; Raguse, J.-D.; Hofmann, T.; et al. A role of the epithelial sodium channel in human salt taste transduction? *Chemosens. Percept.* **2008**, *1*, 78–90. [CrossRef]
49. Huque, T.; Cowart, B.J.; Dankulich-Nagrudny, L.; Pribitkin, E.A.; Bayley, D.L.; Spielman, A.I.; Feldman, R.S.; Mackler, S.A.; Brand, J.G. Sour ageusia in two individuals implicates ion channels of the ASIC and PKD families in human sour taste perception at the anterior tongue. *PLoS ONE* **2009**, *4*, e7347. [CrossRef]
50. Waldmann, R.; Champigny, G.; Bassilana, F.; Voilley, N.; Lazdunski, M. Molecular cloning and functional expression of a novel amiloride-sensitive Na^+ channel. *J. Biol. Chem.* **1995**, *270*, 27411–27414. [CrossRef]
51. Ji, H.-L.; LaToya, R.; Bishop, L.R.; Anderson, S.J.; Catherine, M.; Fuller, C.M.; Benos, D.J. The role of Pre-H2 domains of α- and δ-epithelial Na^+ channels in ion permeation, conductance, and amiloride sensitivity. *J. Biol. Chem.* **2004**, *279*, 8428–8440. [CrossRef]
52. Ji, H.-L.; Su, X.-F.; Kedar, S.; Li, J.; Barbry, P.; Smith, P.R.; Matalon, S.; Benos, D.J. δ-subunit confers novel biophysical features to αβγ-human epithelial sodium channel (ENaC) via a physical interaction. *J. Biol. Chem.* **2006**, *281*, 8233–8241. [CrossRef]
53. Bachmanov, A.A.; Bosak, N.P.; Lin, C.; Matsumoto, I.; Ohmoto, M.; Reed, D.R.; Nelson, T.M. Genetics of taste receptors. *Curr. Pharmaceut. Des.* **2014**, *20*, 2669–2683. [CrossRef] [PubMed]
54. Petrik, D.; Myoga, M.H.; Grade, S.; Gerkau, N.J.; Pusch, M.; Rose, C.R.; Grothe, B.; Götz, M. Epithelial sodium channel regulates adult neural stem cell proliferation in a flow-dependent manner. *Cell Stem Cell* **2018**, *22*, 865–878. [CrossRef] [PubMed]
55. Feng, P.; Huang, L.; Wang, H. Taste bud homeostasis in health, disease, and aging. *Chem. Senses* **2014**, *39*, 3–16. [CrossRef] [PubMed]
56. Barlow, L.A. Progress and renewal in gustation: New insights into taste bud development. *Development* **2015**, *142*, 3620–3629. [CrossRef] [PubMed]

57. Dias, A.G.; Rousseau, D.; Duizer, L.; Cockburn, M.; Chiu, W.; Nielsen, D.; El-Sohemy, A. Genetic variation in putative salt taste receptors and salt taste perception in humans. *Chem. Senses* **2013**, *38*, 137–145. [CrossRef] [PubMed]
58. Firsov, D.; Schild, L.; Gautschi, I.; Mérillat, A.M.; Schneeberger, E.; Rossier, B.C. Cell surface expression of the epithelial Na channel and a mutant causing Liddle syndrome: A quantitative approach. *Proc. Natl. Acad. Sci. USA* **1996**, *93*, 15370–15375. [CrossRef] [PubMed]
59. Konstas, A.A.; Korbmacher, C. The γ-subunit of ENaC is more important for channel surface expression than the β-subunit. *Am. J. Physiol. Cell Physiol.* **2003**, *284*, C447–C456. [CrossRef]
60. Contreras, A.C. Salt taste and disease. *Am. J. Clin. Nutr.* **1978**, *31*, 1088–1097. [CrossRef]
61. Beauchamp, G.K. The human preference for excess salt. *Am. Sci.* **1987**, *75*, 27–33.
62. Duncan, C.J. Salt preference of birds and mammals. *Physiol. Zool.* **1962**, *35*, 120–132. [CrossRef]
63. Oka, Y.; Butnaru, M.; von Buchholtz, L.; Ryba, N.J.; Zuker, C.S. High salt recruits aversive taste pathways. *Nature* **2013**, *494*, 472–475. [CrossRef] [PubMed]
64. Dötsch, M.; Busch, J.; Batenburg, M.; Liem, G.; Tareilus, E.; Mueller, R.; Meijer, G. Strategies to reduce sodium consumption: A food industry perspective. *Crit. Rev. Food Sci. Nutr.* **2009**, *49*, 841–851. [CrossRef] [PubMed]
65. Burnier, M.; Wuerzner, G.; Bochud, M. Salt, blood pressure and cardiovascular risk: What is the most adequate preventive strategy? A Swiss perspective. *Front. Physiol.* **2015**, *6*, 227. [CrossRef] [PubMed]
66. Mozaffarian, D.; Fahimi, S.; Singh, G.M.; Micha, R.; Khatibzadeh, S.; Engell, R.E.; Lim, S.; Danaei, G.; Ezzati, M.; Powles, J. Global Burden of Diseases Nutrition and Chronic Diseases Expert Group. Global sodium consumption and death from cardiovascular causes. *N. Engl. J. Med.* **2014**, *371*, 624–634. [CrossRef] [PubMed]
67. Xu, J.J.; Elkaddi, N.; Garcia-Blanco, A.; Spielman, A.I.; Bachmanov, A.A.; Chung, H.Y.; Ozdener, M.H. Arginyl dipeptides increase the frequency of NaCl-elicited responses via epithelial sodium channel alpha and delta subunits in cultured human fungiform taste papillae cells. *Sci. Rep.* **2017**, *7*, 7483. [CrossRef]
68. Schindler, A.; Dunkel, A.; Stähler, F.; Backes, M.; Ley, J.; Meyerhof, W.; Hofmann, T. Discovery of salt taste enhancing arginyl dipeptides in protein digests and fermented fish sauces by means of a sensomics approach. *J. Agric. Food Chem.* **2011**, *59*, 12578–12588. [CrossRef]
69. Kellenberger, S.; Schild, L. Epithelial sodium channel/degenerin family of ion channels: A variety of functions for a shared structure. *Physiol. Rev.* **2002**, *82*, 735–767. [CrossRef]
70. Ogawa, T.; Nakamura, T.; Tsuji, E.; Miyanaga, Y.; Nakagawa, H.; Hirabayashi, H.; Uchida, T. The combination effect of L-arginine and NaCl on bitterness suppression of amino acid solutions. *Chem. Pharm. Bull.* **2004**, *52*, 172–177. [CrossRef]
71. Carpenter, G.H. The secretion, components, and properties of saliva. *Annu. Rev. Food Sci. Technol.* **2013**, *4*, 267–276. [CrossRef]
72. Stolle, T.; Grondinger, F.; Dunkel, A.; Meng, C.; Médard, G.; Kuster, B.; Hofmann, T. Salivary proteome patterns affecting human salt taste sensitivity. *J. Agric. Food Chem.* **2017**, *65*, 9275–9286. [CrossRef]
73. Hughey, R.P.; Carattino, M.D.; Kleyman, T.R. Role of proteolysis in the activation of epithelial sodium channels. *Curr. Opin. Nephrol. Hypertens.* **2007**, *16*, 444–450. [CrossRef] [PubMed]
74. Kleyman, T.R.; Carattino, M.D.; Hughey, R.P. ENaC at the cutting edge: Regulation of epithelial sodium channels by proteases. *J. Biol. Chem.* **2009**, *284*, 20447–20451. [CrossRef] [PubMed]
75. Stolle, T.; Grondinger, F.; Dunkel, A.; Hofmann, T. Quantitative proteomics and SWATH-MS to elucidate peri-receptor mechanisms in human salt taste sensitivity. *Food Chem.* **2018**, *254*, 95–102. [CrossRef] [PubMed]
76. Li, X.-J.; Blackshaw, S.; Snyder, S.H. Expression and localization of amiloride-sensitive sodium channel indicate a role for non-taste cells in taste perception. *Proc. Natl. Acad. Sci. USA* **1994**, *91*, 1814–1818. [CrossRef]
77. Weisz, O.A.; Johnson, J.P. Noncoordinate regulation of ENaC: Paradigm lost? *Am. J. Physiol. Ren. Physiol.* **2002**, *285*, F833–F842. [CrossRef]

nutrients

MDPI

Article

Expression of Renin-Angiotensin System Components in the Taste Organ of Mice

Noriatsu Shigemura [1,2,*], Shingo Takai [1], Fumie Hirose [1,3], Ryusuke Yoshida [1,4], Keisuke Sanematsu [1,2] and Yuzo Ninomiya [1,2,5,*]

1 Section of Oral Neuroscience, Graduate School of Dental Sciences, Kyushu University, 3–1–1 Maidashi, Higashi-ku, Fukuoka 812–8582, Japan; takai.shingo.774@m.kyushu-u.ac.jp (S.T.); f-hirose@dent.kyushu-u.ac.jp (F.H.); yoshida.ryusuke@okayama-u.ac.jp (R.Y.); sanematu@dent.kyushu-u.ac.jp (K.S.)

2 Division of Sensory Physiology, Development Center for Five-Sense Devices, Kyushu University, Fukuoka 819-0395, Japan

3 Section of Orthodontics and Dentofacial Orthopedics, Division of Oral Health, Growth, and Development, Faculty of Dental Science, Kyushu University, Fukuoka 812–8582, Japan

4 Department of Oral Physiology, Graduate School of Medicine, Dentistry and Pharmaceutical Sciences, Okayama University, Okayama 700-8525, Japan

5 Monell Chemical Senses Center, Philadelphia, PA 19104, USA

* Correspondence: shigemura@dent.kyushu-u.ac.jp (N.S.); ninomiya.yuzo.285@m.kyushu-u.ac.jp (Y.N.); Tel.: +81-92-642-6312 (N.S.); +81-92-642-4613 (Y.N.)

Received: 2 September 2019; Accepted: 15 September 2019; Published: 19 September 2019

check for updates

Abstract: The systemic renin-angiotensin system (RAS) is an important regulator of body fluid and sodium homeostasis. Angiotensin II (AngII) is a key active product of the RAS. We previously revealed that circulating AngII suppresses amiloride-sensitive salt taste responses and enhances the responses to sweet compounds via the AngII type 1 receptor (AT1) expressed in taste cells. However, the molecular mechanisms underlying the modulation of taste function by AngII remain uncharacterized. Here we examined the expression of three RAS components, namely renin, angiotensinogen, and angiotensin-converting enzyme-1 (ACE1), in mouse taste tissues. We found that all three RAS components were present in the taste buds of fungiform and circumvallate papillae and co-expressed with αENaC (epithelial sodium channel α-subunit, a salt taste receptor) or T1R3 (taste receptor type 1 member 3, a sweet taste receptor component). Water-deprived mice exhibited significantly increased levels of renin expression in taste cells ($p < 0.05$). These results indicate the existence of a local RAS in the taste organ and suggest that taste function may be regulated by both locally-produced and circulating AngII. Such integrated modulation of peripheral taste sensitivity by AngII may play an important role in sodium/calorie homeostasis.

Keywords: taste; sodium taste; renin; angiotensin II; angiotensinogen; angiotensin-converting enzyme

1. Introduction

The renin-angiotensin system (RAS) is a major hormone system involved in body fluid and sodium homeostasis [1]. Angiotensin II (AngII), an octapeptide hormone, is the most powerful biologically active product of the RAS and plays important roles in the regulation of vascular tone, cardiac function, and renal sodium re-absorption. AngII is also thought to be a potent stimulator of sodium appetite and preference. For example, intracerebroventricular or intravenous infusion of AngII in the rat produces dose-dependent salt appetite and stimulates sodium intake over a range of concentrations that are normally rejected [2,3]. The gustatory system provides critical information about the quality and nutritional value of food before it is ingested. Thus, changes in sodium taste sensitivity might contribute to the ingestive behaviors induced by AngII. We recently addressed this hypothesis and revealed that AngII suppresses amiloride-sensitive salt taste responses and enhances the responses to sweet compounds via the AngII type 1 receptor (AT1) expressed in taste cells, without any effects on the amiloride-insensitive salt, sour, bitter, or umami responses [4]. These results suggest that the taste organ is a peripheral target of AngII and that AngII may function to increase sodium intake through the specific reduction in amiloride-sensitive salt taste sensitivity and increase energy intake through the enhancement of sweet responses. However, the molecular mechanisms underlying the modulation of taste function by AngII remain uncharacterized.

It is generally known that AngII is produced by the classical or circulating RAS [1]. Renin is a proteolytic enzyme released primarily from the juxtaglomerular cells of the kidney in response to a decrease in arterial blood pressure or sodium chloride level in the nephron [5]. Angiotensinogen is secreted constitutively, mainly by hepatic cells, into the circulation. Renin cleaves angiotensinogen at the N-terminus to form the decapeptide, angiotensin I (AngI). AngI is converted to AngII through the removal of two C-terminal residues by circulating angiotensin-converting enzyme-1 (ACE1), which is found in various organs, including the lung and kidney [6]. AT1, the main receptor subtype for circulating AngII, is widely distributed throughout the body including vascular smooth muscle, kidney, heart, brain, and taste organ [4,7]. AngII is degraded into smaller active peptides, AngIII, AngIV, and Ang (1–7), by endopeptidases or carboxypeptidases such as ACE2, which is a homolog of ACE1 [8].

In addition to the circulating RAS, it is now recognized that tissues such as the kidneys, brain, heart, adrenal glands and vasculature each have an organ-specific RAS [9,10]. For example, all the RAS components are present in the kidneys, and intrarenal AngII is produced independently of the circulating RAS to function as a paracrine factor via AT1. Inappropriate activation of the intrarenal RAS contributes to the pathogenesis of hypertension and renal injury [11,12]. RAS components are also present in cardiac myocytes and fibroblasts, where they synthesize AngII intracellularly [13,14]. Hyperglycemia selectively upregulates the intracellular RAS system in cardiac myocytes and vascular smooth muscle cells [15], and this is associated with cardiomyocyte apoptosis, oxidative stress, and fibrosis in diabetic rats [16]. The above findings raise the possibility that an organ-specific RAS might also exist in taste tissues and that AngII might be produced locally in response to changes in the peripheral oral environment (e.g., a change in the sodium concentration or osmolality of the saliva or the presence of chemical compounds in foods).

To explore this possibility, we utilized reverse transcription-polymerase chain reaction (RT-PCR), in situ hybridization, and double-staining immunohistochemistry to investigate the expression patterns of renin, angiotensinogen, and ACE in the taste tissues of mice under hydrated and dehydrated conditions.

2. Materials and Methods

All experimental procedures were performed in accordance with the Guidelines for the Care and Use of Laboratory Animals established by the National Institutes of Health and approved by the Committee for Laboratory Animal Care and Use at Kyushu University, Japan. The ethical approval code for the animal experiments was A27-009.

2.1. Animals

This study used male and female C57BL/6NCrj mice (B6; Charles River, Tokyo, Japan), taste receptor type 1 member 3 (T1R3)-green fluorescent protein (GFP) mice [17] and glutamate decarboxylase 67 (GAD67)-GFP mice [18] aged 8–16 weeks of age and weighing 21–30 g. The mice were housed at a constant temperature (24 ± 1 °C) under a 12 h–12 h light-dark cycle (lights on at 08:00) and given access to food and water ad libitum.

2.2. Reverse Transcription-Polymerase Chain Reaction (RT-PCR)

RT-PCR was performed as described previously [4,19–21]. Mouse taste buds in the peeled epithelium were individually removed from fungiform or circumvallate papillae by aspiration with a transfer pipette. The RNeasy Plus Micro kit (Qiagen, Stanford, CA, USA) was used to purify RNAs from 100 fungiform or circumvallate taste buds from three mice or from a 1 mm × 1 mm block of epithelial tissue without taste buds. cDNAs were synthesized by RT [oligo(dT)12–18 primer] using the SuperScript pre-amplification system (Invitrogen, Carlsbad, CA, USA). Two protocols were used to prevent genomic DNA from contributing to the signal: (1) primers were chosen to span one or more introns to distinguish the PCR products from genomic DNA; The primer sequences are shown in Supplementary Table S1, and (2) RNA was handled in parallel in the presence and absence of reverse transcriptase. PCR was performed using the following conditions: 95 °C for 5 min (one cycle); 94 °C for 15 s, 58 °C for 30 s, 68 °C for 40–80 s (25–40 cycles); and 75 °C for 5 min (one cycle). Each 20 µL of PCR solution contained 0.5 U of Taq DNA polymerase (TaKaRa Ex TaqHS; Takara Bio, Kusatsu, Japan), 2 µL of 10× PCR buffer containing 20 mmol/L Mg^{2+}, 0.2 mmol/L of each deoxyribonucleotide triphosphate, and 0.6 µmol/L of each primer pair. The resulting amplification products were visualized in a 2% agarose gel with 0.5 µg/mL ethidium bromide.

2.3. In Situ Hybridization

In situ hybridization experiments were performed as described previously [4,21–24]. PCR products amplified using specific primer pairs for RAS genes (Supplementary Table S1) were purified and cloned into the pGEM T-easy vector (Promega, Madison, WI, USA) and confirmed by direct sequencing. Digoxigenin-labeled antisense RNA probes were synthesized by in vitro transcription using the digoxigenin-ribonucleic acid (DIG-RNA) Labeling Mix and T7 or SP6 RNA polymerase (Roche, Mannheim, Germany). Frozen blocks of the dissected anterior parts of the tongue embedded in optimum-cutting temperature (OCT) compound (Sakura Fine technical, Tokyo, Japan) were sectioned into 6-µm thick slices, which were mounted on silane-coated glass slides. The cryosections were fixed in 4% paraformaldehyde in phosphate-buffered saline (PBS) for 10 min at room temperature, treated two times with 0.1% diethylpyrocarbonate in PBS for 15 min, washed with 5× saline sodium citrate buffer (SSC) for 15 min at room temperature, and then prehybridized in a hybridization buffer consisting of 50% formamide, 5× SSC, 5× Denhardt's solution, 500 µg/mL denatured salmon testis DNA and 250 µg/mL denatured baker's yeast tRNA for 1 h at room temperature. Hybridization was performed for 18 h at 58 °C in a hybridization buffer that included 200 ng/mL antisense (or sense) RNA probe. After hybridization, the sections were washed two times in 5× SSC for 5 min each and two times in 0.2× SSC for 30 min each at 65 °C. Subsequently, the sections were immersed in Tris-buffered saline (TBS) consisting of 50 mmol/L Tris-HCl (pH 7.5), and 150 mmol/L NaCl for 5 min at room temperature, treated with a blocking solution containing 0.5% blocking reagent (Roche) in TBS for 30 min, and applied with anti-digoxigenin Fab fragments-conjugated to alkaline phosphatase (AP; 1:400 dilution; Roche) in blocking solution for 60 min at room temperature. After three washes of 5 min each in Tris-NaCl-Tween 20 (TNT) buffer consisting of 50 mmol/L Tris-HCl (pH 7.5), 150 mmol/L NaCl and 0.05% Tween 20, the sections were immersed in AP buffer comprising 100 mmol/L Tris-HCl (pH 9.5), 100 mmol/L NaCl, and 50 mmol/L $MgCl_2$ for 5 min. The signals were developed using 5-bromo-4-chloro-3-indolylphosphate and nitroblue-tetrazolium chloride as chromogenic substrates. Next, the reaction was stopped by washing the slides with Tris-ethylenediaminetetraacetic acid (EDTA) buffer, after which the slides were mounted. The signal specificity of the mRNA for each gene in the taste tissues was tested using a sense RNA probe as a negative control.

2.4. Immunohistochemistry

Immunohistochemistry was performed as described previously [4,20]. The dissected tongues of B6, T1R3-GFP or GAD67-GFP mice were fixed in 4% paraformaldehyde in PBS for 45 min at 4 °C. For immunohistochemical analyses for rennin, mice were deprived of water for 47 h (23 h water deprivation, 1 h water drinking, and 23 h water deprivation) before tongue dissection. After dehydration with sucrose solution (10% for 1 h, 20% for 1 h, and 30% for 3 h, at 4 °C), the tongue frozen block was embedded in OCT compound (Sakura Fine technical) and sectioned into 8-µm thick slices, which were mounted on silane-coated glass slides and air-dried. The sections were rinsed with TNT buffer, exposed to 1% blocking reagent (Roche) for 1 h at room temperature, and applied overnight at 4 °C with primary antibodies targeting renin (1:100 dilution; sc-27318; Santa Cruz Biotechnology, Santa Cruz, CA, USA), angiotensinogen (1:100; 11992-1-AP; Protein tech, Chicago, IL, USA), ACE1 (1:100; sc-12187; Santa Cruz Biotechnology), ecto-nucleoside triphosphate diphosphohydrolase-2 (ENTPDase2; 1:100; AF5797; R&D Systems, Minneapolis, MN, USA), epithelial sodium channel α-subunit (αENaC; 1:100; AB3530P; Millipore, Darmstadt, Germany), AT1 (1:100, sc-1173; Santa Cruz Biotechnology) or Gα-gustducin (gustducin; 1:100; sc-395; Santa Cruz Biotechnology) in 1% blocking reagent. After washing with TNT buffer, the tissues were incubated for 2 h at room temperature with peroxidase or alkaline phosphatase-conjugated secondary antibodies (1:500–1000; Jackson ImmunoResearch Laboratories, Philadelphia, PA, USA) in 1% blocking reagent, and this was followed by incubation for 30 min at room temperature with tyramide-Alexa 568 (for GFP-mice) or tyramide-Alexa 488 substrates (TSA kit; Invitrogen) for the detection of renin, ACE1 or angiotensinogen. After washing with TNT, the tissues were incubated with AP buffer for 5 min at room temperature followed by HNPP/FastRed AP substrate (HNPP fluorescent detection kit; Roche) for 40 min at room temperature to detect the signals of the counterpart. The immunofluorescence of labeled cells and GFP fluorescence were observed using a confocal laser scanning microscope (Fluoview FV-1000; Olympus Corp., Tokyo, Japan) and accompanying software. Nomarski images were also obtained in order to visualize individual cells in the taste buds.

To evaluate the number of cells expressing renin, angiotensinogen, ACE1, T1R3-GFP, GAD67-GFP, αENaC, ENTPDase2 and AT1, Nomarski images were overlaid with immunofluorescence (or GFP) images, we then counted the number of positive cells displaying apparent apical processes and/or perinuclear region in each taste bud in horizontal sections of fungiform papillae and circumvallate papillae. Image-Pro Plus v4.0 (Media Cybernetics, Rockville, MD, USA) was used to exclude artifactual signals: cells were considered positive if their signal density was greater than the mean plus two standard deviations (SDs) of that of taste cells in the negative control (primary antibodies omitted). The same cells found on contiguous sections were counted only once.

2.5. *Quantitative Densitometric Analysis*

To examine whether the expression of renin, a principal initiator of the RAS cascade, in taste cells was upregulated in response to dehydration, B6 mice were deprived of water for 47 h (23 h water deprivation, 1 h water drinking, and 23 h water deprivation). The mice had free access to food throughout the procedure. Dissected tongues from water-deprived and non-water-deprived mice ($n = 3$, each) were fixed at the same time in 4% paraformaldehyde in PBS for 45 min at 4 °C. After dehydration, embedding, and sectioning at 8 μm, both tongue slices were mounted on a single silane-coated glass slide. Then, renin signals were detected by single-molecule immunohistochemistry using CF568-conjugated anti-goat IgG (1:200, Biotium, Hayward, CA, USA) as a secondary antibody instead of a tyramide amplification system, in order to avoid excessive catalyzed reporter deposition of tyramide. All procedures, including image capture, were performed under the same conditions (incubation periods, temperatures, reagent volumes/concentrations, and exposure values for image capture). The taste cell in each section was delineated by comparison with adjacent sections counterstained with gustducin (a type II taste cell marker proposed to be a bitter taste-related G-protein in mouse circumvallate papillae) [25].

Measurement of the total area of a positive cell, and the mean signal intensity was performed using Adobe Photoshop CS6 (Adobe Systems, San Jose, CA, USA) [26]. Regions of interest were drawn around immuno-positive cells in the captured images, and the mean total area and the mean and median pixel intensities of the signals were determined using the Measurement tool in Photoshop. The pixel intensities were normalized against the background by subtracting the value obtained from an area outlying the positive cells in the same section. Two sections of the circumvallate papillae at regular intervals were analyzed for each mouse.

2.6. Quantitative PCR (qPCR)

To further examine whether the expression of renin in taste cells was upregulated in response to dehydration, the relative abundance of renin mRNAs in circumvallate taste papillae was examined by using a quantitative PCR method as previously described [24]. The isolated taste buds from each circumvallate papillae of each mouse were pooled (n = 4–7 water-deprived mice, n = 3–7 non-water-deprived mice). As a positive control for renin expression, a 1 mm^3 block of the kidney was also collected from each mouse (n = 7 and 7, water-deprived and non-water-deprived mice, respectively). Total RNA extraction was performed as described in RT-PCR section. The RNA concentration was measured using NanoDrop ND-1000 (Thermo Fisher Scientific, Waltham, MA, USA). The purity of nucleic acids was assessed by calculating the A260/A280 absorbance ratio. SuperScript VILO Master Mix (cat. no. 11755050, Thermo Fisher Scientific) was used for cDNA synthesis. For quantitative real-time PCR, Fast SYBR Green Master Mix (Applied Biosystems, CA, USA) was used. PCR was performed as follows: 95 °C for 20 s (one cycle); 95 °C for 3 s, 60 °C for 30 s (40 cycles); and 95 °C for 15 s, 60 °C for 1 min, 95 °C for 15 s (one cycle for melting curve analysis) using the ABI StepOnePlus system (Applied Biosystems). Data were analyzed with the StepOne Software (ver. 2.3, Applied Biosystems). The presence of a single amplicon was verified by melting curve analysis, and by agarose gel electrophoresis. Data were obtained from at least three independent experiments, and all reactions were run in triplicate. The quantitative PCR data were normalized using the ΔΔCt method with GAPDH (glyceraldehyde-3-phosphate dehydrogenase) in each sample as reference. ΔΔCt values were calculated by subtracting the average ΔCt of the non-water-deprived samples from each ΔCt of both the non-water-deprived and the water-deprived samples. Fold change from non-water-deprived to water-deprived conditions was calculated as $2^{(-\Delta\Delta CT)}$. All primer pairs for renin, gustducin, keratin 8, and GAPDH were chosen such that the primers are in separate exons. The PCR primers used for each gene are presented in Supplementary Table S1.

2.7. Statistical Analysis

All values are given as the mean ± standard error of the mean. The data were statistically analyzed using Student's t-test for unpaired samples (Excel; Microsoft Corp., Redmond, WA, USA). $p < 0.05$ was considered statistically significant.

3. Results

3.1. RT-PCR Reveals Renin-Angiotensin-Related Gene Expression in Mouse Taste Buds

The expressions of renin, angiotensinogen, ACE1, and ACE2 mRNAs in the taste cells of B6 mice were examined by RT-PCR. As shown in Figure 1A, bands of the correct size (460 bp for renin, 318 bp for angiotensinogen, 398 bp for ACE1, and 303 bp for ACE2) were evident in taste papillae. Renin, angiotensinogen, and ACE1 mRNAs were expressed in fungiform and circumvallate papillae but not in tongue epithelium devoid of taste buds. Similarly, RT-PCR products for a taste cell marker, transient receptor potential channel M5 (TRPM5; 368 bp) [27–30], were also found in fungiform and circumvallate papillae but not in tongue epithelium devoid of taste buds. ACE2 mRNA was expressed in all tissues. As a positive control, β-actin mRNA (360 bp) was also detected in all tissues. All control experiments in which the reverse transcriptase enzyme was omitted (RT-) yielded negative results.

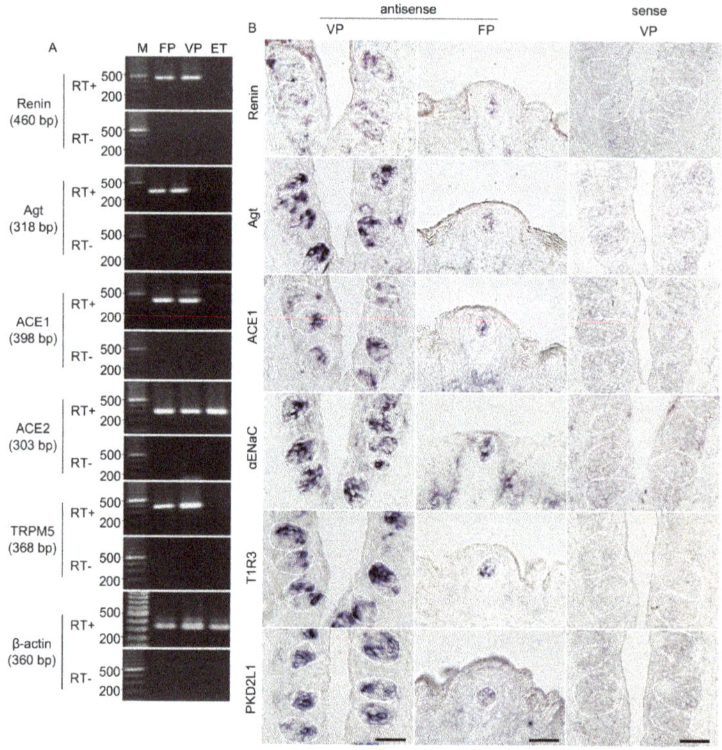

Figure 1. Renin, angiotensinogen (Agt) and angiotensin-converting enzyme-1 (ACE1) mRNAs are expressed in mouse taste bud cells. (*A*) Reverse transcription-polymerase chain reaction (RT-PCR) amplification of renin, Agt, ACE1, ACE2, transient receptor potential channel M5 (TRPM5), and β-actin mRNAs from fungiform papillae (FP), circumvallate papillae (VP), and tongue epithelium devoid of taste cells (ET). RT+ and RT− conditions are, respectively, with and without reverse transcriptase. M (bp): 100 bp marker ladder. (**B**) In situ hybridization detection of renin, Agt, ACE1, epithelial sodium channel α subunit (αENaC), taste receptor type 1 member 3 (T1R3), and polycystic kidney disease 2-like 1 (PKD2L1) in FP and VP of B6 mice. The sense probes served as a negative control. Dotted lines indicate the outlines of taste buds: scale bar, 50 μm.

3.2. Renin-Angiotensin-Related Genes Are Localized to a Subset of Taste Bud Cells

In situ hybridization experiments detected renin, angiotensinogen, and ACE1 mRNA in a subset of cells in the fungiform and circumvallate papillae of mice but not in surrounding epithelial cells (Figure 1B). Comparable results were obtained for three markers of taste cells: αENaC (an amiloride-sensitive salt taste receptor subunit candidate) [31], T1R3 (a sweet/umami taste receptor component) [32,33] and polycystic kidney disease 2-like 1 (PKD2L1; a sour taste-related molecule) [34,35] (Figure 1B). Control hybridizations using sense probes for renin, angiotensinogen, ACE1, αENaC, T1R3, and PKD2L1 were negative. These results, together with the RT-PCR data, strongly suggest that renin, angiotensinogen, and ACE1 are expressed in mouse taste bud cells of both the anterior and posterior tongue.

3.3. Renin-Angiotensin-Related Proteins Are Co-Expressed with T1R3 or αENaC in Taste Bud Cells

Immunohistochemistry experiments detected renin, ACE1, and angiotensinogen in some spindle-shaped taste cells of the fungiform and circumvallate papillae of the mouse but not in surrounding tissues or gustatory nerves. Notably, co-expression of renin and taste cell markers was observed (Figure 2 and Table 1). In both the fungiform and circumvallate papillae, a subset of renin-positive cells expressed αENaC (ENaC/renin: 83.9% in fungiform papillae and 88.1% in circumvallate papillae) and T1R3 (as marked by T1R3-GFP; T1R3/renin: 54.7% in fungiform papillae and 49.2% in circumvallate papillae). Renin-expressing cells also showed immunoreactivity for AT1 (AT1/renin: 72.0% in fungiform papillae and 80.0% in circumvallate papillae) [4], but renin expression was not found in sour/type III cells as marked by GAD67-GFP (GAD/renin: 0% in both fungiform and circumvallate papillae).

The co-expression of ACE1 and taste cell markers was also examined (Figure 3 and Table 2). ACE1-positive cells expressed αENaC (ENaC/ACE1: 71.9% in fungiform papillae and 91.3% in circumvallate papillae) and T1R3 (T1R3/ACE1: 70.0% in fungiform papillae and 57.5% in circumvallate papillae). ACE1-expressing cells also exhibited positivity for AT1 (AT1/ACE1: 81.3% in fungiform papillae and 79.7% in circumvallate papillae). ACE1 expression was rarely observed in GAD67-expressing cells (GAD/ACE1: 0% in fungiform papillae and 2.7% in circumvallate papillae).

Figure 4 and Table 3 present data regarding the co-expression of angiotensinogen and taste cell markers. The majority of angiotensinogen-positive cells expressed T1R3 (T1R3/angiotensinogen: 73.0% in fungiform papillae and 70.7% in circumvallate papillae) and renin (renin/angiotensinogen: 75.8% in fungiform papillae and 70.6% in circumvallate papillae). A subset of cells expressing ENTPDase2, a type I taste cell marker [36], showed positive signals for angiotensinogen (ENTPDase2/angiotensinogen: 37.5% in fungiform papillae and 34.1% in circumvallate papillae). Angiotensinogen was not observed in GAD67-expressing cells (GAD/angiotensinogen: 0% in both fungiform and circumvallate papillae).

The inverse co-expression ratios (renin, ACE1, and angiotensinogen/taste cell markers) are shown in Tables 1–3. A summary of the expression patterns of renin, ACE1, angiotensinogen, and taste cell markers in the fungiform and circumvallate papillae is shown in Figure 5.

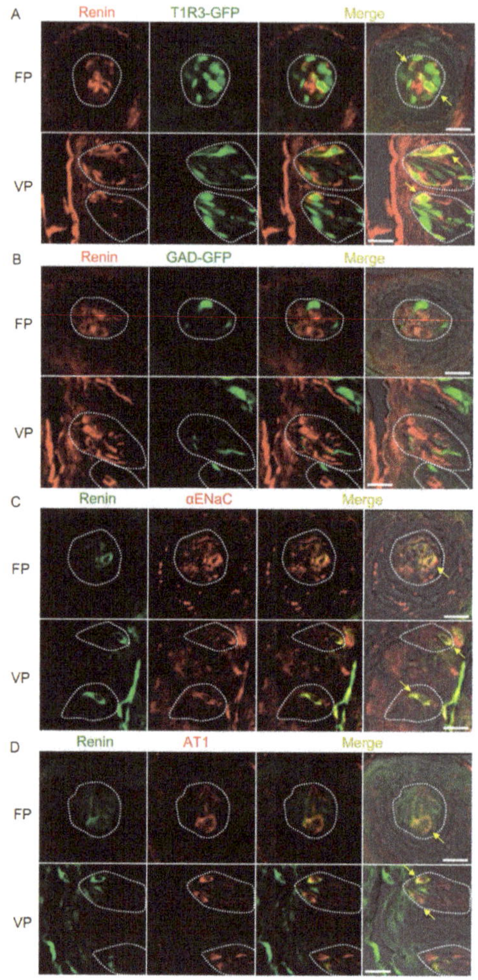

Figure 2. Co-expression of renin with taste receptor family 1 member 3 (T1R3), glutamate decarboxylase 67 (GAD), epithelial sodium channel α subunit (αENaC) or angiotensin II type 1 receptor (AT1) in taste bud cells. (**A**) Expression of renin in fungiform papillae (FP) and circumvallate papillae (VP) of T1R3-green fluorescent protein (GFP) mice. (**B**) Expression of renin in FP and VP of GAD67-GFP mice. (**C**) Co-expression of renin with αENaC in FP and VP of B6 mice. (**D**) Co-expression of renin with AT1 in FP and VP of B6 mice. Immunostaining for renin is shown in red (**A**,**B**) or green (**C**,**D**). GFP fluorescence in T1R3-GFP and GAD67-GFP mice is shown in green (**A**,**B**). Immunostaining for αENaC and AT1 is shown in red (**C**,**D**). The last merge panels: Nomarski images were overlaid with immunofluorescence (or GFP) merge images. Arrows indicate renin-expressing taste cells that co-express T1R3, αENaC, or AT1. Dotted lines indicate the outlines of taste buds: scale bar, 25 μm.

Table 1. Co-expression ratio of Renin and taste cell markers in fungiform (FP) and circumvallate papillae (VP) in mice.

	FP		VP			FP		VP	
T1R3/Renin	54.7%	(29/53, n = 12)	49.2%	(31/63, n = 17)	Renin/T1R3	39.7%	(29/73, n = 12)	36.0%	(31/86, n = 17)
GAD/Renin	0%	(0/33, n = 8)	0%	(0/42, n = 11)	Renin/GAD	0%	(0/13, n = 8)	0%	(0/33, n = 11)
ENaC/Renin	83.9%	(26/31, n = 10)	88.1%	(37/42, n = 11)	Renin/ENaC	92.9%	(26/28, n = 10)	86.0%	(37/43, n = 11)
AT1/Renin	72.0%	(18/25, n = 9)	80.0%	(32/40, n = 11)	Renin/AT1	72.0%	(18/25, n = 9)	91.4%	(32/35, n = 11)

AT1: angiotensin II type 1 receptor; ENaC: epithelial sodium channel α-subunit; GAD: glutamate decarboxylase; T1R3: taste receptor type 1 member 3. The number of protein1 + protein2 doubly labeled cell/number of protein2 positive cell and n = number of taste buds examined are shown in parentheses.

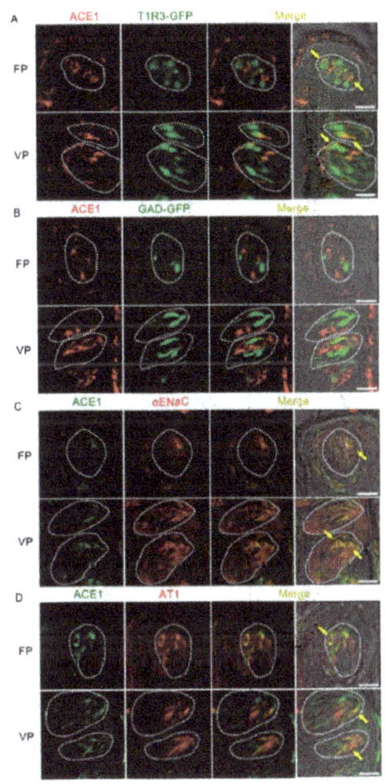

Figure 3. Co-expression of angiotensin-converting enzyme-1 (ACE1) with taste receptor type 1 member 3 (T1R3), glutamate decarboxylase 67 (GAD), epithelial sodium channel α subunit (αENaC) or angiotensin II type 1 receptor (AT1) in taste bud cells. (**A**) Expression of ACE1 in fungiform papillae (FP) and circumvallate papillae (VP) of T1R3-green fluorescent protein (GFP) mice. (**B**) Expression of ACE1 in FP and VP of GAD67-GFP mice. (**C**) Co-expression of ACE1 with αENaC in FP and VP of B6 mice. (**D**) Co-expression of ACE1 with AT1 in FP and VP of B6 mice. Immunostaining for ACE1 is shown in red (**A**,**B**) or green (**C**,**D**). GFP fluorescence in T1R3-GFP and GAD67-GFP mice is shown in green (**A**,**B**). Immunostaining for αENaC and AT1 is shown in red (**C**,**D**). The last merge panels: Nomarski images were overlaid with immunofluorescence (or GFP) merge images. Arrows indicate ACE1-expressing taste cells that co-express T1R3, αENaC, or AT1. Dotted lines indicate the outlines of taste buds: scale bar, 25 μm.

Table 2. Co-expression ratio of ACE1 and taste cell markers in fungiform (FP) and circumvallate papillae (VP) in mice.

	FP		VP			FP		VP	
T1R3/ACE	70.0%	(35/50, n = 11)	57.5%	(42/73, n = 20)	ACE/T1R3	58.3%	(35/60, n = 11)	46.2%	(42/91, n = 20)
GAD/ACE	0%	(0/38, n = 11)	2.7%	(2/75, n = 16)	ACE/GAD	0%	(0/15, n = 11)	4.3%	(2/47, n = 16)
ENaC/ACE	71.9%	(23/32, n = 7)	91.3%	(42/46, n = 10)	ACE/ENaC	85.2%	(23/27, n = 7)	93.3%	(42/45, n = 10)
AT1/ACE	81.3%	(26/32, n = 7)	79.7%	(63/79, n = 13)	ACE/AT1	96.3%	(26/27, n = 7)	98.4%	(63/64, n = 13)

ACE1: angiotensin-converting enzyme-1; AT1: angiotensin II type 1 receptor; ENaC: epithelial sodium channel α-subunit; GAD: glutamate decarboxylase; T1R3: taste receptor type 1 member 3. The number of protein1 + protein2 doubly labeled cell/number of protein2 positive cell, and n = number of taste buds examined are shown in parentheses.

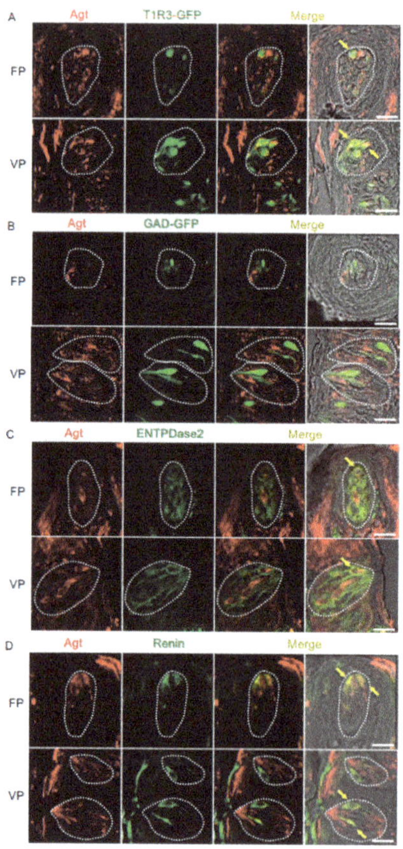

Figure 4. Co-expression of angiotensinogen (Agt) with taste receptor type 1 member 3 (T1R3), glutamate decarboxylase 67 (GAD), nucleoside triphosphate diphosphohydrolase-2 (ENTPDase2) and renin in taste bud cells. (**A**) Expression of Agt in fungiform papillae (FP) and circumvallate papillae (VP) of T1R3-green fluorescent protein (GFP) mice. (**B**) Expression of Agt in FP and VP of GAD67-GFP mice. (**C**) Co-expression of Agt with ENTPDase2 in FP and VP of B6 mice. (**D**) Co-expression of Agt with renin in FP and VP of B6 mice. Immunostaining for Agt is shown in red. GFP fluorescence in T1R3-GFP and GAD67-GFP mice and immunostaining for ENTPDase2 and renin are shown in green. The last merge panels: Nomarski images were overlaid with immunofluorescence (or GFP) merge images. Arrows indicate Agt-expressing taste cells that co-express T1R3, ENTPDase2, or renin. Dotted lines indicate the outlines of taste buds: scale bar, 25 μm.

Table 3. Co-expression ratio of Agt and taste cell markers in fungiform (FP) and circumvallate papillae (VP) in mice.

	FP		VP			FP		VP	
T1R3/Agt	73.0%	(27/37, n = 7)	70.7%	(41/58, n = 14)	Agt/T1R3	60.0%	(27/45, n = 7)	37.6%	(41/109, n = 14)
GAD/Agt	0%	(0/10, n = 4)	0%	(0/56, n = 17)	Agt/GAD	0%	(0/8, n = 4)	0%	(0/68, n = 17)
ENTPD2/Agt	37.5%	(9/24, n = 9)	34.1%	(15/44, n = 14)	Agt/ENTPD2	8.7%	(9/103, n = 9)	7.9%	(15/189, n = 14)
Renin/Agt	75.8%	(25/33, n = 9)	70.6%	(24/34, n = 9)	Agt/Renin	92.6%	(25/27, n = 9)	85.7%	(24/28, n = 9)

Agt: angiotensinogen; ENTPDase2: ecto-nucleoside triphosphate diphosphohydrolase-2; GAD: glutamate decarboxylase; T1R3: taste receptor type 1 member 3. The number of protein1 + protein2 doubly labeled cell/number of protein2 positive cell, and n = number of taste buds examined are shown in parentheses.

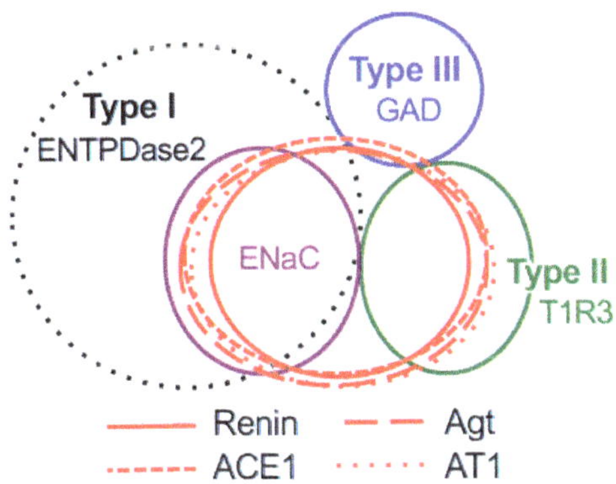

Figure 5. Summary of the patterns of co-expression between renin, angiotensinogen (Agt), angiotensin-converting enzyme-1 (ACE1), angiotensin II type 1 receptor (AT1) and type-specific markers of taste cells [nucleoside triphosphate diphosphohydrolase-2 (ENTPDase2) for cell type I, taste receptor type 1 member 3 (T1R3) for cell type II, and glutamate decarboxylase 67 (GAD) for cell type III] in both the fungiform and circumvallate papillae of mice.

3.4. Renin Expression Is Upregulated in the Taste Buds Cells of Water-Deprived Mice

To examine whether water deprivation leads to changes in the abundance of renin, we performed immunohistochemistry and quantitative image analyses using circumvallate papillae from non-deprived and water-deprived mice. In non-deprived mice (used as the control), immunoreactivity for renin was observed in the apical process regions of taste cells with weaker expression in the cell body regions (Figure 6A). In water-deprived mice, immunoreactivity for renin was observed throughout some cells, i.e., from apical to basal regions (Figure 6B). There were no significant differences between non-deprived and water-deprived mice in the mean number of positive cells per bud [2.75 ± 0.2 (n = 28) vs. 2.85 ± 0.2 (n = 27)] or the mean area (pixels) of a positive cell [576.7 ± 25.0 (n = 28) vs. 563.2 ± 27.4 (n = 28)] (Figure 6C,D). However, significant differences were observed between control and water-deprived mice in both the mean pixel intensity (arbitrary units) [14.5 ± 0.9 (n = 28) vs. 19.2 ± 1.5 (n = 28), $p < 0.05$] and the median pixel intensity (arbitrary units) [10.3 ± 0.6 (n = 28) vs. 15.4 ± 1.4 (n = 28), $p < 0.01$] of the immuno-positive cells (Figure 6E,F).

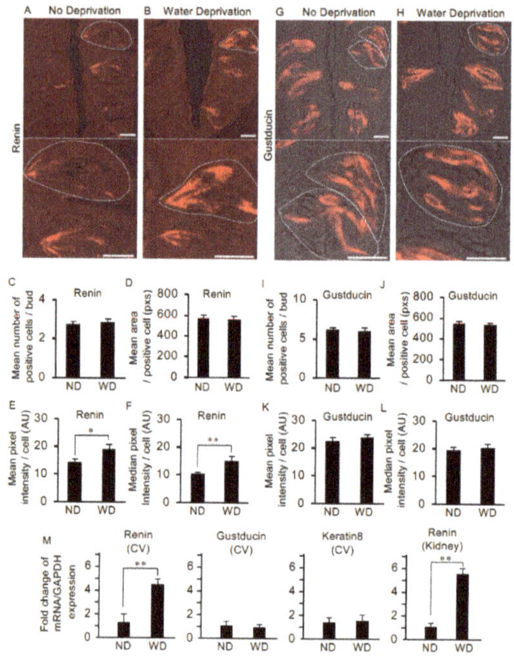

Figure 6. The level of renin expression in taste cells is upregulated after water deprivation. (**A**) In non-deprived (control) mice, renin was expressed at the apical process regions of taste cells and more weakly in the cell body regions (lower image: higher magnification of the upper image). (**B**) In mice deprived of water for 47 h, renin expression was observed throughout some cells, i.e., from apical to basal regions. Water deprivation did not change the mean number of renin-positive cells per taste bud [n = 28 taste buds from three non-water deprived mice (ND), n = 27 taste buds from three water-deprived mice (WD)] (**C**) or the mean area of renin-positive cells (n = 28 taste cells from ND, n = 28 taste cells from WD) (pixels: pxs) (**D**). However, significant differences were detected in the mean pixel intensity (**E**) and median pixel intensity (**F**) of renin-positive cells (arbitrary unit: AU). Gustducin immunoreactivity was detected throughout positive cells from the apical to the basal regions in both non-water-deprived mice (**G**) and water-deprived mice (**H**). Water deprivation did not change the mean number of gustducin-positive cells per taste bud (n = 23 taste buds from ND, n = 23 taste buds from WD) (**I**), mean area of gustducin-positive cells (n = 32 taste cells from ND, n = 38 taste cells from WD) (pixels: pxs) (**J**), mean pixel intensity of gustducin-positive cells (**K**) or median pixel intensity (**L**) of gustducin-positive cells (arbitrary unit: AU). (**M**) Quantification of mRNA expression of renin, gustducin and keratin 8 by real-time PCR in the circumvallate taste papilla (CV) and Kidney (as the control tissue for renin expression) from ND and WD. Data were obtained from at least three independent experiments (n = 3–7 mice) per group each PCR assays were performed in triplicate. The quantitative PCR results were normalized using the ΔΔCt method with GAPDH (glyceraldehyde-3-phosphate dehydrogenase) in each sample as the reference, and shown as fold change of mRNA expression compared to ND. * $p < 0.05$, ** $p < 0.01$ for comparisons between ND and WD groups (Student's t-test). All data are presented as the mean ± standard error of the mean. Dotted lines indicate the outlines of taste buds: scale bar, 25 μm.

We also examined differences in the abundance of gustducin protein (a taste-specific G-protein mostly expressed in bitter taste cells in mouse circumvallate papillae) in taste cells between control and water-deprived mice. Immunoreactivity for gustducin was observed throughout the apical-to-basal regions of a subset of taste cells in both the non-deprived and water-deprived groups (Figure 6G,H). We observed no significant differences between the non-deprived and water-deprived groups in the mean number of positive cells per taste bud [6.13 ± 0.3 (n = 23) vs. 6.09 ± 0.3 (n = 23)], the mean area of a positive cell [548.5 ± 20.4 (n = 32) vs. 532.8 ± 20.5 (n = 38)], the mean pixel intensity of a positive cell [22.8 ± 1.1 (n = 32) vs. 23.8 ± 1.2 (n = 38)] or the median pixel intensity of a positive cell [19.6 ± 1.1 (n = 32) vs. 20.6 ± 1.2 (n = 38)] (Figure 6I–L).

Next, we asked whether renin mRNA expression in circumvallate taste bud cells is upregulated after water-deprivation by using quantitative PCR. mRNA levels of genes were normalized for GAPDH as the endogenous reference gene and shown as fold change of mRNA expression compared to non-water-deprived groups (Figure 6M). Statistical analysis by *t*-test revealed a significant difference in relative renin mRNA expression in circumvallate taste buds between control and water-deprived mice [1.36 ± 0.62 (n = 3 mice) vs. 4.60 ± 0.46 (n = 4), $p < 0.01$]. In the Kidney as the control tissue for renin expression, renin mRNA was significantly increased in the water-deprived group compared with the non-water-deprived group [1.11 ± 0.22 (n = 7) vs. 5.53 ± 0.51 (n = 7), $p < 0.01$) (Figure 6M). We also performed quantitative PCR analysis of gustducin and keratin 8 (a pan-taste cell marker) in the circumvallate taste bud cells, and observed no significant differences in the mRNA expression between control and water-deprived groups [gustducin: 1.14 ± 0.33 (n = 5) vs. 1.02 ± 0.20 (n = 7), keratin 8: 1.37 ± 0.43 (n = 6) vs. 1.51 ± 0.52 (n = 6)] (Figure 6M). Together, these results suggest that renin expression is upregulated in taste cells in response to dehydration.

4. Discussion

In the present study, we found that three RAS components, namely renin, ACE1, and angiotensinogen, were present in the taste buds of fungiform and circumvallate papillae and co-expressed with αENaC (a salt taste receptor), T1R3 (a sweet taste receptor component) [31–33] and AT1 [4]. Furthermore, significantly increased levels of renin expression were observed in taste cells after water deprivation of the mouse. These results suggest that the taste organ contains a local RAS that may be capable of producing AngII within the taste buds. AngII may function as a modulator of amiloride-sensitive salt and sweet taste sensitivities via AT1 in an autocrine and paracrine manner.

In animals with AngII-dependent hypertension, the intrarenal AngII level is higher than that which can be explained on the basis of equilibration with circulating AngII [12]. This suggests that AngII can be synthesized within the kidney. Indeed, all components of the RAS are present in the distal nephron [11,37]. Principal cells in the connecting tubule and the cortical collecting duct abundantly express renin, a key initiator of the RAS, which cleaves the N-terminal end of angiotensinogen to generate AngI in the tubular fluid. Renin immunolocalizes predominately to the apical side of the cytoplasm of principal cells. Quantitative histological analysis revealed that mice exhibited minimal renin immuno-staining (below the detection level) in the connecting tubule under high sodium diets, in contrast, overnight sodium restriction led to a marked increase in the number of the renin-positive connecting tubule cells. Furthermore, the amount of renin mRNA was shown to increase in response to sodium restriction [11,37]. In our study, renin immunostaining predominated at the apical regions of taste cells, which face the oral cavity and are available to sense various chemical compounds in saliva or foods (Figure 6A). Moreover, significantly increased levels of renin in taste cells were observed in water-deprived mice (Figure 6E,F,M), which may be consistent with the results observed in connecting tubule cells in the kidney [11]. These results indicate that renin synthesis initiating the RAS cascade occurs in taste cells and may be regulated in response to factors in the oral environment such as sodium concentration (osmolality) in the saliva or chemical compounds in foods, which are sensed at the apical regions of the taste cells.

A previous study showed that amiloride-sensitive NaCl taste responses were suppressed 10–30 min after intraperitoneal injection of AngII, suggesting that systemic AngII produced by the circulating RAS is able to modulate NaCl taste responses in taste cells [4]. Thus, there may be two pathways regulating taste, namely the circulating RAS and the local RAS. What would be the advantage of the co-existence of these two RASs? One possibility is that the local RAS assists in maintaining the constancy of body NaCl balance through acute changes in AngII production in response to fasting or random perturbations in the oral cavity during feeding and drinking. In other words, the local RAS could help to avert decreases in body sodium levels or prevent excessive sodium/calorie consumption independently of changes in body NaCl balance. However, once NaCl balance has been changed, the distal nephron NaCl concentration would serve to regulate the circulating RAS, which supports the maintenance of prolonged deviations of body NaCl concentration from the normal set point [5]. Such temporally integrated regulation of taste sensitivity by local AngII (short-term feedforward regulation predicting changes in body fluid composition) and systemic AngII (long-term negative feedback regulation in response to changes in body salt/water balance) may play an important role in sodium/energy homeostasis. ACE2, which degrades Ang II to Ang (1–7) to oppose the actions of Ang II, is present not only in fungiform and circumvallate papillae but also in the tongue epithelium (Figure 1A). This observation suggests that AngII generated locally in taste buds can be rapidly degraded by ACE2, which would support the hypothesis of short-term regulation of taste sensitivity by a local RAS.

Multiple lines of evidence from studies using molecular approaches indicate that the basic taste qualities (sweet, salty, bitter, sour, and umami) are mediated by distinct taste cells expressing unique taste receptors [38]. Sweet, umami and bitter substances activate G-protein coupled receptors (T1R2+T1R3 for sweet [32,33], T1R1+T1R3 for umami [33] and T2Rs for bitter [39]) and subsequent common signaling pathways involving phospholipase C-β [28], the type 3 inositol 1,4,5-trisphosphate receptor [40] and TRPM5 [27–30]. Salty and sour substances are believed to activate channel-type receptors (ENaC for amiloride-sensitive salt taste [31,41] and PKD2L1/1L3 for sour taste [34,35]). Each of these taste receptors is expressed in a different set of taste cells [38]. The present study demonstrated that all three of the RAS components studied are co-expressed with αENaC or T1R3 but not GAD67, suggesting that taste-regulated AngII production by the local RAS may be mediated by αENaC-expressing, amiloride-sensitive salt taste cells, and T1R3-expressing sweet taste cells. The expression patterns of the RAS components in the taste organ may relate to previous observations that AngII suppresses amiloride-sensitive salt taste responses and enhances sweet taste responses without any effects on bitter, umami and sour responses [4]. The relationship between salt and sweet preferences via AngII signaling, analogous to that between reduced salt-sensitive neural responses and increased sugar-sensitive neural responses to dietary NaCl in the rat nucleus of the solitary tract after intracerebroventricular renin infusion on chronic deoxycorticosterone acetate (DOCA, the precursor of aldosterone) treatment [42], or that between sodium and glucose absorption via sodium-glucose cotransporter-1 in the small intestine [43], may optimize sodium and calorie intake via the taste system.

The macula densa is a group of 15–20 epithelial cells in the distal convoluted tubule of the kidney. These cells play a critical role in sensing changes in tubular fluid composition and sending signals to the juxtaglomerular apparatus that controls renin release [44]. It has been shown that increasing the NaCl concentration at the macula densa suppresses renin release, whereas reducing the NaCl concentration results in a prompt stimulation of renin release [45]. NaCl sensing by the macula densa involves apical NaCl transport mechanisms, including the furosemide-sensitive $Na^+/K^+/2Cl^-$cotransporter (NKCC), which is the primary NaCl entry mechanism [46,47]. The apical membranes of macula densa cells also express Na^+/H^+ exchanger (NHE), which participates in Na^+ transport and the regulation of intracellular pH and cell volume [48,49]. It has been reported that NHE1 and NHE3 were detected in taste receptor cells [50]. It is possible that NKCC or NHE may participate in sensing changes in NaCl concentration at the cell membrane of taste cells.

It is generally accepted that an elevation in the circulating level of AngII inhibits renin secretion from the juxtaglomerular apparatus of the kidney. However, intrarenal AngII increases renin mRNA and protein levels in the distal nephron [51], indicating positive-feedback regulation of intrarenal RAS by AngII. In experiments involving renal cross-transplantation between global AT1-knockout mice and wild-type controls, AngII was shown to cause hypertension through stimulation of AT1 receptors in the kidney [52]. Overexpression of renin in the collecting duct caused spontaneous hypertension [53]. The Aldosterone/NaCl-induced RAS functional impairment also, not only caused a reduction of the salt taste sensitivity, but also salt-sensitive hypertension in rat [54]. These results suggest that intrarenal RAS contributes to the pathogenesis of hypertension. The existence of such a positive-feedback mechanism in the taste organ potentially would have broad implications, since continuous activation of the taste organ RAS might be associated with hypertension induced by excessive salt consumption through sustained low salt taste sensitivity.

5. Conclusions

We have demonstrated that the taste organ has three major components of the RAS, namely renin, angiotensinogen, and ACE1, which would enable AngII to be produced locally in the taste buds. Expression analyses showed that the RAS components are co-expressed with αENaC or T1R3 in a subset of taste cells. Renin immunoreactivity was detected at the apical regions of taste cells, which face the oral cavity and thus are exposed to various chemical compounds in saliva or food. Furthermore, renin synthesis in taste tissue was significantly upregulated in response to water deprivation. Taken together, these results suggest the existence of a previously unidentified local RAS in the taste organ. The specific reduction of amiloride-sensitive salt taste sensitivity and enhancement of sweet taste sensitivity may be mediated by both locally-produced AngII (temporal feedforward regulation) and circulating AngII (continuous negative feedback regulation). Such an integrated regulation of peripheral taste sensitivity by AngII may play an important role in sodium/calorie homeostasis.

Supplementary Materials: The following are available online at http://www.mdpi.com/2072-6643/11/9/2251/s1, Table S1: PCR primers.

Author Contributions: Conceptualization—N.S.; Methodology—N.S.; Validation—S.T. and K.S.; Investigation—N.S., S.T., F.H., R.Y. and K.S.; Data curation—N.S., S.T., F.H., K.S. and R.Y.; Writing—Original Draft Preparation—N.S.; Writing—Review & Editing—N.S. and Y.N.

Funding: This research was supported in part by Grants-in-Aid for Scientific Research 15K11044, 18K19652 and 19H03818 (N.S.), 18077004 and 18109013 (Y.N.) from the Ministry of Education, Culture, Sports, Science and Technology of Japan. The authors thank OXMEDCOMMS (www.oxmedcomms.com) for writing assistance.

Acknowledgments: The authors thank S. Damak (Nestlé Research Center, Lausanne, Switzerland) and R.F. Margolskee (Monell Chemical Senses Center, Philadelphia, PA, USA) for providing the original stock of T1R3-GFP mice and Y. Yanagawa (Gunma University Graduate School of Medicine, Gunma, Japan) for providing the original stock of GAD67-GFP mice.

Conflicts of Interest: The authors declare no conflict of interest.

References

1. Fyhrquist, F.; Saijonmaa, O. Renin-angiotensin system revisited. *J. Intern. Med.* **2008**, *264*, 224–236. [CrossRef] [PubMed]
2. Avrith, D.B.; Fitzsimons, J.T. Increased sodium appetite in the rat induced by intracranial administration of components of the renin–angiotensin system. *J. Physiol.* **1980**, *301*, 349–364. [CrossRef] [PubMed]
3. Fitts, D.A.; Thunhorst, R.L. Rapid elicitation of salt appetite by an intravenous infusion of angiotensin II in rats. *Am. J. Physiol.* **1996**, *270*, R1092–R1098. [CrossRef] [PubMed]
4. Shigemura, N.; Iwata, S.; Yasumatsu, K.; Ohkuri, T.; Horio, N.; Sanematsu, K.; Yoshida, R.; Margolskee, R.F.; Ninomiya, Y. Angiotensin II modulates salty and sweet taste sensitivities. *J. Neurosci.* **2013**, *33*, 6267–6277. [CrossRef] [PubMed]
5. Braun-Menéndez, E.; Fasciolo, J.C.; Leloir, L.F.; Munoz, J.M. La sustancia hipertensora de la sangre del riñón isquemido. *Rev. Soc. Argent. Biol.* **1939**, *15*, 420–425.
6. Skeggs, L.T., Jr.; Khan, J.R.; Shumway, N.P. The preparation and function of the hypertensin-converting enzyme. *J. Exp. Med.* **1956**, *103*, 295–299. [CrossRef] [PubMed]
7. De Gasparo, M.; Catt, K.J.; Inagami, T.; Wright, J.W.; Unger, T. International Union of Pharmacology. XXIII. The angiotension II receptors. *Pharmacol. Rev.* **2000**, *52*, 415–472.
8. Donoghue, M.; Hsieh, F.; Baronas, E.; Godbout, K.; Gosselin, M.; Stagliano, N.; Donovan, M.; Woolf, B.; Robinson, K.; Jeyaseelan, R.; et al. A novel angiotensin-converting enzyme-related carboxypeptidase (ACE2) converts angiotensin I to angiotensin 1–9. *Circ. Res.* **2000**, *87*, E1–E9. [CrossRef] [PubMed]

9. Fleming, I.; Kohlstedt, K.; Busse, R. The tissue renin-angiotensin system and intracellular signalling. *Curr. Opin. Nephrol. Hypertens.* **2006**, *15*, 8–13. [CrossRef]
10. Paul, M.; Poyan Mehr, A.; Kreutz, R. Physiology of local renin-angiotensin systems. *Physiol. Rev.* **2006**, *86*, 747–803. [CrossRef]
11. Rohrwasser, A.; Morgan, T.; Dillon, H.F.; Zhao, L.; Callaway, C.W.; Hillas, E.; Zhang, S.; Cheng, T.; Inagami, T.; Ward, K.; et al. Elements of a paracrine tubular renin-angiotensin system along the entire nephron. *Hypertension* **1999**, *34*, 1265–1274. [CrossRef] [PubMed]
12. Navar, L.G.; Prieto, M.C.; Satou, R.; Kobori, H. Intrarenal angiotensin II and its contribution to the genesis of chronic hypertension. *Curr. Opin. Pharmacol.* **2011**, *11*, 180–186. [CrossRef] [PubMed]
13. Dostal, D.E.; Baker, K.M. The cardiac renin-angiotensin system: Conceptual, or a regulator of cardiac function? *Circ. Res.* **1999**, *85*, 643–650. [CrossRef] [PubMed]
14. Kumar, R.; Yong, Q.C.; Thomas, C.M.; Baker, K.M. Intracardiac intracellular angiotensin system in diabetes. *Am. J. Physiol. Regul. Integr. Comp. Physiol.* **2012**, *302*, R510–R517. [CrossRef] [PubMed]
15. Haller, H.; Lindschau, C.; Erdmann, B.; Quass, P.; Luft, F.C. Effects of intracellular angiotensin II in vascular smooth muscle cells. *Circ. Res.* **1996**, *79*, 765–772. [CrossRef] [PubMed]
16. Singh, V.P.; Le, B.; Khode, R.; Baker, K.M.; Kumar, R. Intracellular angiotensin II production in diabetic rats is correlated with cardiomyocyte apoptosis, oxidative stress, and cardiac fibrosis. *Diabetes* **2008**, *57*, 3297–3306. [CrossRef] [PubMed]
17. Damak, S.; Mosinger, B.; Margolskee, R.F. Transsynaptic transport of wheat germ agglutinin expressed in a subset of type II taste cells of transgenic mice. *BMC Neurosci.* **2008**, *9*, 96. [CrossRef] [PubMed]
18. Tamamaki, N.; Yanagawa, Y.; Tomioka, R.; Miyazaki, J.; Obata, K.; Kaneko, T. Green fluorescent protein expression and colocalization with calretinin, parvalbumin, and somatostatin in the GAD67-GFP knock-in mouse. *J. Comp. Neurol.* **2003**, *467*, 60–79. [CrossRef] [PubMed]
19. Kawai, K.; Sugimoto, K.; Nakashima, K.; Miura, H.; Ninomiya, Y. Leptin as a modulator of sweet taste sensitivities in mice. *Proc. Natl. Acad. Sci. USA* **2000**, *97*, 11044–11049. [CrossRef]
20. Yoshida, R.; Ohkuri, T.; Jyotaki, M.; Yasuo, T.; Horio, N.; Yasumatsu, K.; Sanematsu, K.; Shigemura, N.; Yamamoto, T.; Margolskee, R.F.; et al. Endocannabinoids selectively enhance sweet taste. *Proc. Natl. Acad. Sci. USA* **2010**, *107*, 935–939. [CrossRef]
21. Shigemura, N.; Ohta, R.; Kusakabe, Y.; Miura, H.; Hino, A.; Koyano, K.; Nakashima, K.; Ninomiya, Y. Leptin modulates behavioral responses to sweet substances by influencing peripheral taste structures. *Endocrinology* **2004**, *145*, 839–847. [CrossRef]
22. Shigemura, N.; Miura, H.; Kusakabe, Y.; Hino, A.; Ninomiya, Y. Expression of leptin receptor (Ob-R) isoforms and signal transducers and activators of transcription (STATs) mRNAs in the mouse taste buds. *Arch. Histol. Cytol.* **2003**, *66*, 253–260. [CrossRef]
23. Shigemura, N.; Islam, A.A.; Sadamitsu, C.; Yoshida, R.; Yasumatsu, K.; Ninomiya, Y. Expression of amiloride-sensitive epithelial sodium channels in mouse taste cells after chorda tympani nerve crush. *Chem. Senses* **2005**, *30*, 531–538. [CrossRef]
24. Shigemura, N.; Ohkuri, T.; Sadamitsu, C.; Yasumatsu, K.; Yoshida, R.; Beauchamp, G.K.; Bachmanov, A.A.; Ninomiya, Y. Amiloride-sensitive NaCl taste responses are associated with genetic variation of ENaC alpha-subunit in mice. *Am. J. Physiol. Regul. Integr. Comp. Physiol.* **2008**, *294*, R66–R75. [CrossRef]
25. Wong, G.T.; Gannon, K.S.; Margolskee, R.F. Transduction of bitter and sweet taste by gustducin. *Nature* **1996**, *381*, 796–800. [CrossRef] [PubMed]
26. Mattila, J.T.; Ojo, O.O.; Kepka-Lenhart, D.; Marino, S.; Kim, J.H.; Eum, S.Y.; Via, L.E.; Barry, C.E., 3rd; Klein, E.; Kirschner, D.E.; et al. Microenvironments in tuberculous granulomas are delineated by distinct populations of macrophage subsets and expression of nitric oxide synthase and arginase isoforms. *J. Immunol.* **2013**, *191*, 773–784. [CrossRef] [PubMed]
27. Pérez, C.A.; Huang, L.; Rong, M.; Kozak, J.A.; Preuss, A.K.; Zhang, H.; Max, M.; Margolskee, R.F. A transient receptor potential channel expressed in taste receptor cells. *Nat. Neurosci.* **2002**, *5*, 1169–1176. [CrossRef]
28. Zhang, Y.; Hoon, M.A.; Chandrashekar, J.; Mueller, K.L.; Cook, B.; Wu, D.; Zuker, C.S.; Ryba, N.J. Coding of sweet, bitter, and umami tastes: Different receptor cells sharing similar signaling pathways. *Cell* **2003**, *112*, 293–301. [CrossRef]

29. Talavera, K.; Yasumatsu, K.; Voets, T.; Droogmans, G.; Shigemura, N.; Ninomiya, Y.; Margolskee, R.F.; Nilius, B. Heat activation of TRPM5 underlies thermal sensitivity of sweet taste. *Nature* **2005**, *438*, 1022–1025. [CrossRef]
30. Damak, S.; Rong, M.; Yasumatsu, K.; Kokrashvili, Z.; Pérez, C.A.; Shigemura, N.; Yoshida, R.; Mosinger, B., Jr.; Glendinning, J.I.; Ninomiya, Y.; et al. Trpm5 null mice respond to bitter, sweet, and umami compounds. *Chem. Senses* **2006**, *31*, 253–264. [CrossRef] [PubMed]
31. Chandrashekar, J.; Kuhn, C.; Oka, Y.; Yarmolinsky, D.A.; Hummler, E.; Ryba, N.J.; Zuker, C.S. The cells and peripheral representation of sodium taste in mice. *Nature* **2010**, *464*, 297–301. [CrossRef] [PubMed]
32. Nelson, G.; Hoon, M.A.; Chandrashekar, J.; Zhang, Y.; Ryba, N.J.; Zuker, C.S. Mammalian sweet taste receptors. *Cell* **2001**, *106*, 381–390. [CrossRef]
33. Li, X.; Staszewski, L.; Xu, H.; Durick, K.; Zoller, M.; Adler, E. Human receptors for sweet and umami taste. *Proc. Natl. Acad. Sci. USA* **2002**, *99*, 4692–4696. [CrossRef] [PubMed]
34. Ishimaru, Y.; Inada, H.; Kubota, M.; Zhuang, H.; Tominaga, M.; Matsunami, H. Transient receptor potential family members PKD1L3 and PKD2L1 form a candidate sour taste receptor. *Proc. Natl. Acad. Sci. USA* **2006**, *103*, 12569–12574. [CrossRef] [PubMed]
35. Huang, A.L.; Chen, X.; Hoon, M.A.; Chandrashekar, J.; Guo, W.; Tränkner, D.; Ryba, N.J.; Zuker, C.S. The cells and logic for mammalian sour taste detection. *Nature* **2006**, *442*, 934–938. [CrossRef] [PubMed]
36. Vandenbeuch, A.; Anderson, C.B.; Parnes, J.; Enjyoji, K.; Robson, S.C.; Finger, T.E.; Kinnamon, S.C. Role of the ectonucleotidase NTPDase2 in taste bud function. *Proc. Natl. Acad. Sci. USA* **2013**, *110*, 14789–14794. [CrossRef] [PubMed]
37. Komlosi, P.; Fuson, A.L.; Fintha, A.; Peti-Peterdi, J.; Rosivall, L.; Warnock, D.G.; Bell, P.D. Angiotensin I conversion to angiotensin II stimulates cortical collecting duct sodium transport. *Hypertension* **2003**, *42*, 195–199. [CrossRef] [PubMed]
38. Chandrashekar, J.; Hoon, M.A.; Ryba, N.J.; Zuker, C.S. The receptors and cells for mammalian taste. *Nature* **2006**, *444*, 288–294. [CrossRef]
39. Chandrashekar, J.; Mueller, K.L.; Hoon, M.A.; Adler, E.; Feng, L.; Guo, W.; Zuker, C.S.; Ryba, N.J. T2Rs function as bitter taste receptors. *Cell* **2000**, *100*, 703–711. [CrossRef]
40. Hisatsune, C.; Yasumatsu, K.; Takahashi-Iwanaga, H.; Ogawa, N.; Kuroda, Y.; Yoshida, R.; Ninomiya, Y.; Mikoshiba, K. Abnormal taste perception in mice lacking the type 3 inositol 1,4,5-trisphosphate receptor. *J. Biol. Chem.* **2007**, *282*, 37225–37231. [CrossRef]
41. Heck, G.L.; Mierson, S.; DeSimone, J.A. Salt taste transduction occurs through an amiloride-sensitive sodium transport pathway. *Science* **1984**, *223*, 403–405. [CrossRef] [PubMed]
42. McCaughey, S.A.; Scott, T.R. Rapid induction of sodium appetite modifies taste-evoked activity in the rat nucleus of the solitary tract. *Am. J. Physiol. Regul. Integr. Comp. Physiol.* **2000**, *279*, R1121–R1131. [CrossRef] [PubMed]
43. Ferraris, R.P.; Diamond, J.M. Crypt/villus site of substrate-dependent regulation of mouse intestinal glucose transporters. *Proc. Natl. Acad. Sci. USA* **1993**, *90*, 5868–5872. [CrossRef]
44. Bell, P.D.; Lapointe, J.Y.; Peti-Peterdi, J. Macula densa cell signaling. *Annu. Rev. Physiol.* **2003**, *65*, 481–500. [CrossRef] [PubMed]
45. Skøtt, O.; Briggs, J.P. Direct demonstration of macula densa-mediated renin secretion. *Science* **1987**, *237*, 1618–1620. [CrossRef] [PubMed]
46. Lorenz, J.N.; Weihprecht, H.; Schnermann, J.; Skøtt, O.; Briggs, J.P. Renin release from isolated juxtaglomerular apparatus depends on macula densa chloride transport. *Am. J. Physiol.* **1991**, *260*, F486–F493. [CrossRef] [PubMed]
47. He, X.R.; Greenberg, S.G.; Briggs, J.P.; Schnermann, J. Effects of furosemide and verapamil on the NaCl dependency of macula densa-mediated renin secretion. *Hypertension* **1995**, *26*, 137–142. [CrossRef]
48. Peti-Peterdi, J.; Chambrey, R.; Bebok, Z.; Biemesderfer, D.; St John, P.L.; Abrahamson, D.R.; Warnock, D.G.; Bell, P.D. Macula densa Na(+)/H(+) exchange activities mediated by apical NHE2 and basolateral NHE4 isoforms. *Am. J. Physiol. Renal Physiol.* **2000**, *278*, F452–F463. [CrossRef]
49. Hanner, F.; Chambrey, R.; Bourgeois, S.; Meer, E.; Mucsi, I.; Rosivall, L.; Shull, G.E.; Lorenz, J.N.; Eladari, D.; Peti-Peterdi, J. Increased renal renin content in mice lacking the Na+/H+ exchanger NHE2. *Am. J. Physiol. Renal Physiol.* **2008**, *294*, F937–F944. [CrossRef]

50. Vinnikova, A.K.; Alam, R.I.; Malik, S.A.; Ereso, G.L.; Feldman, G.M.; McCarty, J.M.; Knepper, M.A.; Heck, G.L.; DeSimone, J.A.; Lyall, V. Na+-H+ exchange activity in taste receptor cells. *J. Neurophysiol.* **2004**, *91*, 1297–1313. [CrossRef]
51. Prieto-Carrasquero, M.C.; Harrison-Bernard, L.M.; Kobori, H.; Ozawa, Y.; Hering-Smith, K.S.; Hamm, L.L.; Navar, L.G. Enhancement of collecting duct renin in angiotensin II-dependent hypertensive rats. *Hypertension* **2004**, *44*, 223–229. [CrossRef]
52. Crowley, S.D.; Gurley, S.B.; Herrera, M.J.; Ruiz, P.; Griffiths, R.; Kumar, A.P.; Kim, H.S.; Smithies, O.; Le, T.H.; Coffman, T.M. Angiotensin II causes hypertension and cardiac hypertrophy through its receptors in the kidney. *Proc. Natl. Acad. Sci. USA* **2006**, *103*, 17985–17990. [CrossRef] [PubMed]
53. Ramkumar, N.; Ying, J.; Stuart, D.; Kohan, D.E. Overexpression of renin in the collecting duct causes elevated blood pressure. *Am. J. Hypertens.* **2013**, *26*, 965–972. [CrossRef] [PubMed]
54. Sakamoto, T.; Fujii, A.; Saito, N.; Kondo, H.; Ohuchi, A. Alteration of amiloride-sensitive salt taste nerve responses in aldosterone/NaCl-induced hypertensive rats. *Neurosci. Res.* **2016**, *108*, 60–66. [CrossRef] [PubMed]

nutrients

MDPI

Article

Exploring Associations between Interindividual Differences in Taste Perception, Oral Microbiota Composition, and Reported Food Intake

Camilla Cattaneo, Patrizia Riso *, Monica Laureati, Giorgio Gargari and Ella Pagliarini

Department of Food, Environmental and Nutritional Sciences (DeFENS), University of Milan, 20133 Milan, Italy; camilla.cattaneo@unimi.it (C.C.); monica.laureati@unimi.it (M.L.); giorgio.gargari@unimi.it (G.G.); ella.pagliarini@unimi.it (E.P.)

* Correspondence: patrizia.riso@unimi.it; Tel.: +39-025-031-6726

Received: 25 April 2019; Accepted: 21 May 2019; Published: 24 May 2019

check for updates

Abstract: The role of taste perception, its relationship with oral microbiota composition, and their putative link with eating habits and food intake were the focus of the present study. A sample of 59 reportedly healthy adults (27 male, 32 female; age: 23.3 ± 2.6 years) were recruited for the study and taste thresholds for basic tastes, food intake, and oral microbiota composition were evaluated. Differences in taste perception were associated with different habitual food consumption (i.e., frequency) and actual intake. Subjects who were orally hyposensitive to salty taste reported consuming more bakery and salty baked products, saturated-fat-rich products, and soft drinks than hypersensitive subjects. Subjects hyposensitive to sweet taste reported consuming more frequently sweets and desserts than the hypersensitive group. Moreover, subjects hypersensitive to bitter taste showed higher total energy and carbohydrate intakes compared to those who perceived the solution as less bitter. Some bacterial taxa on tongue dorsum were associated with gustatory functions and with vegetable-rich (e.g., Prevotella) or protein/fat-rich diets (e.g., Clostridia). Future studies will be pivotal to confirm the hypothesis and the potential exploitation of oral microbiome as biomarker of long-term consumption of healthy or unhealthy diets.

Keywords: taste sensitivity; taste thresholds; food records; food intake; oral microbiota; eating habits

1. Introduction

There are many known drivers of food choice and habits, however, taste is considered one of the main predictors [1]. It is generally assumed that humans perceive five different taste modalities: bitter, sweet, umami, sour, and salty. Each taste quality is associated with different nutritional or physiological requirements, or indicates a potential dietary risk [2]. Sweet, salty, and umami are supposed to signal the nutrient composition of foods, with sweet taste representing carbohydrates, salty taste associated with electrolytes, and umami with proteins. On the contrary, stimuli categorized as bitter and sour are associated with compounds that could be potentially harmful, and are generally regarded as innate aversions [3,4]. Taste perception varies greatly among individuals, strongly influencing food preferences and selection, and therefore nutritional status and health [5]. In particular, during the last decades, research has been focusing on bitter taste perception, and the genetic predisposition to perceive the bitter taste of 6-n-propylthiouracil (PROP) has gained considerable attention as a prototypical taste stimulus and an oral marker of food preferences and eating behavior [6]. Some additional markers include the density of fungiform papillae on the tongue tip [7] and thermal tasting [8].

Previous studies suggested a connection between individuals' taste sensitivity and food acceptance and consumption [9–11]. It has been conclusively demonstrated that PROP-sensitive individuals detect more bitterness from glucosinolate-containing vegetables than non-sensitive individuals and an association between variation in bitter taste perception and food preferences has been documented [12–15]. However, the potential interaction between bitterness sensitivity and food intake has yet to be fully understood [10,16]. Furthermore, it has been reported that variation in sweet taste perception between individuals (see [17] for a review) could influence food selection and overall dietary intake. However, even if a number of previous studies [18–23] have investigated this relationship, conflicting results have been reported, probably due to differences in study participants' characteristics (e.g., gender, ethnicity, age), in sweet taste perception assessment (e.g., psychophysical measurement, type of sweet stimuli) or in dietary intake assessment (e.g., food record, food frequency questionnaire).

The variability in response to salty stimuli has been examined for decades, but a direct genetic link to human salt taste perception has yet to be discovered [16]. The relationship between salty sensitivity and food intake has been studied much less but a connection between individuals' salt taste sensitivity and sodium-rich foods acceptance and consumption has been suggested [18,19]. Moreover, salty sensitivity appears to be determined more by environmental factors, including exposure to NaCl and consumption of specific nutrients, than by heritability components [24–28].

Recently, in addition to the study of the perception of basic tastes, increasing attention has been focused on the sensitivity to fat stimulus since various evidence indicated that humans can perceive non-esterified, long-chain fatty acids in the oral cavity [29,30]. Moreover, it has been suggested that fat perception may influence the choice and consumption of some high-fat foods and, thus, possibly affects body weight [31,32]. However, additional studies are needed to confirm this assumption.

In this context, the potentiality of nongenetic factors to interact with genetic predisposition and influence food habits should be adequately considered. Recently, it has been suggested that differences in oral microbiota could be involved in the interindividual differences in taste perception. Indeed, in agreement with other reports [33,34], we previously reported a relationship between reduced taste perception and specific oral bacteria's growth [35].

The impact of taste sensitivity and its putative influence on food intake were the focus of the present study. We explored whether variation in gustatory functions among individuals could be related to different dietary patterns and intake. Moreover, gustatory functions and dietary patterns were studied in relation to oral microbiota composition.

2. Material and Methods

2.1. Participants

Healthy, normal-weight volunteers, 18–30 years of age, were recruited from the University of Milan community through public advertisement. They received oral and written explanations of the protocol and answered questionnaires aimed at applying exclusion criteria. The following exclusion criteria were considered: (1) smokers; (2) pregnant or lactating women; (3) subjects on medication that may interfere with their ability to taste; (4) history of food allergies that may interfere with the study evaluations; (5) subjects on antibiotics two months before the study. Participants were asked to refrain from eating, drinking (except room temperature water), brushing teeth, and chewing gum for 3 h prior to testing.

Power analysis was conducted to determine an appropriate sample size to achieve adequate power. Using data from previous studies [17,20,21] and an α of 5% and a β of 10% (90% power), it was calculated that a sample size of 51 would be required to classify at least 20% of the subjects as hypersensitive to all basic taste.

Informed, written consent was obtained from all subjects. The present study was performed according to the principles established by the Declaration of Helsinki and the protocol was approved by the Institutional Ethics Committee of the University of Milan (protocol number 16/17).

2.2. Gustatory Function Assessments

Seven concentrations for each taste stimulus were prepared to determine the recognition thresholds. These concentration ranges covered the published threshold values [36–38], and were adjusted according to preliminary tests. Concentration ranges were established such that the lowest concentration was clearly below and the highest concentration was clearly above the level at which subjects could detect or recognize the stimulus, and allowed for interindividual threshold differences. The dilution factor and the final concentration range were reported in Table 1.

Table 1. Concentrations (g/L) of sucrose, sodium chloride, caffeine, and citric acid used to determine recognition thresholds.

Taste Quality	Reference Stimuli	Sample Concentration (g/L) [a]						
		1	2	3	4	5	6	7
Sweet	Sucrose	1.6×10^{-1}	4.0×10^{-1}	1.02	2.56	6.4	16.0	40.0
Salty	Sodium chloride	6.25×10^{-2}	1.25×10^{-2}	2.5×10^{-1}	5.0×10^{-1}	1.0	2.0	4.0
Bitter	Caffeine	3.0×10^{-3}	9.0×10^{-3}	3.0×10^{-2}	8.0×10^{-2}	2.4×10^{-1}	8.0×10^{-1}	2.0
Sour	Citric acid	2.0×10^{-2}	5.0×10^{-2}	8.0×10^{-2}	1.5×10^{-1}	3.5×10^{-1}	7.5×10^{-1}	1.5

[a] The concentration series for sucrose, sodium chloride, and caffeine were prepared with successive 0.4 log dilution steps. The concentration series for citric acid were prepared with successive 0.3 log dilution steps. Reference chemical details: sucrose (Sigma Aldrich srl, Milano, Italy), sodium chloride (Sigma Aldrich srl, Milano, Italy), caffeine (Sigma Aldrich srl, Milano, Italy), and citric acid (Sigma Aldrich srl, Milano, Italy).

For each taste, participants received the samples at each concentration as a three-alternative forced-choice (3-AFC) ascending series, according to ISO/DIS 13301:2018 [39]. Starting from the lowest concentration, three samples (10 mL each), one containing the sample with stimulus and distilled water and two background samples (only with distilled water), were presented at each concentration. Participants were asked to take the whole 10 mL of sample into their mouth, swirl the solution around for 3 s, and expectorate. Using the forced-choice method, participants were instructed to select, or guess, the sample which was different from the other two. All samples were given a three-digit number, and the position of the samples with stimuli was randomly allocated. Participants were asked to rinse their mouth with distilled water between each concentration step.

2.3. Food Intake Evaluation

A Food and Beverage Frequency Questionnaire (FB-FFQ) was used to assess the consumption frequency of specific categories of foods and beverages over the previous month. The questionnaire was developed by considering a previous validated questionnaire for the Italian population [40] but specifically focusing on the main important food and beverage classes contributing to identify "consumers' behavior according to taste sensitivity" more than actual energy and nutrient intake. In fact, the main purpose was to assess in a qualitative way the habitual intake of foods and beverages. Participants indicated their frequency of intake of 22 food categories (i.e., sweets, salty snacks, dairy products, meats, fish, fruit, vegetables) using the following frequency categories: less than once a month, 1–3 times per month, 1–4 times per week, 5–7 times per week, 2-4 times per day, and 5 or more times per day [20,41].

In addition, each participant completed a seven-day food diary to assess their food and nutrient intake. Participants were given verbal instructions and written examples on how to fill in the diary recording the type and amount of foods consumed and possibly the recipes and method of preparation. All participants were given a food record booklet.

2.4. Oral Sample Collection, DNA Extraction, and Microbiota Composition Evaluation

Oral sample collection was performed as previously reported in Cattaneo et al. [35]. In brief, volunteers, sitting in front of a mirror, were asked to protrude the tongue and gently press the swab on the surface, rolling and touching edges, tip, and all defined areas of the tongue (~2/3 of tongue length) for 2 min using a sterile flocked swab (FLOQSwabsTM, COPAN S.p.A., Brescia, Italy). The swab samples were immediately placed in 750 μL of Power Bead solution provided in the DNeasy PowerLyzer PowerSoil DNA extraction kit (Qiagen, Hilden, Germany) and stored at −80 °C. For DNA extraction, samples were thawed on ice, homogenized for five minutes, and dried. Then, samples were processed using the DNA extraction kit and following manufacturer's instructions with a minor modification (e.g., samples were incubated at 65 °C for 10 min after adding C1 solution). Bacterial cell disruption was performed mechanically using a Precellys bead beater kept in a cold room (3 cycles of 6800 rpm × 30 s; Advanced Biotech Italia s.r.l., Seveso, Italy). Quantification and verification of the 260/280 ratio of the extracted DNA was carried out with a Take3 Micro-Volume plate in a Gen5 microplate reader (BioTek Instrument Inc., Winooski, VT, USA). Finally, DNA samples were stored at −80 °C. The bacterial taxonomic composition of oral swabs was assessed in a previous study by 16S rRNA gene profiling using Illumina HiSeq technology [35]. Analysis and taxonomic assignment of sequencing reads were also performed by means of the bioinformatic pipeline Quantitative Insights Into Microbial Ecology (QIIME) version 1.9.0 with the GreenGenes database (version 13_5). Metadata were deposited in the European Nucleotide Archive (ENA) of the European Bioinformatics Institute under accession code PRJEB28769.

2.5. Data Analysis

The matrix of the correct and incorrect answers produced separately by each judge was used to calculate the individual thresholds. The individual's Best Estimate Threshold (BET) for each sensory stimulus was calculated as the geometric mean of the highest concentration missed and the next highest concentration that was correctly recognized (ISO/DIS 13301:2018) [39].

Participants were divided according to their taste sensitivity into three groups, using basic taste thresholds as a grouping variable. Participants were defined as hypersensitive if they presented threshold values in the lower percentile (25th percentile): Salty ≤ 0.088 g/L; Sweet ≤ 0.639 g/L; Bitter ≤ 0.0164 g/L; Sour ≤ 0.0316 g/L, and as hyposensitive if they presented threshold values in the higher percentile (75th percentile): Salty ≥ 0.0353 g/L; Sweet ≥ 4.040 g/L; Bitter ≥ 0.1385 g/L; Sour ≥ 0.1095 g/L. The remaining subjects were considered as medium sensitive.

Food record data were used to estimate total energy and macronutrient intake by using the software MètaDieta developed using Italian food composition tables (METEDA srl, Italy).

The FB-FFQ data registered for the 22 food item categories were converted to daily frequency equivalents (DFE) calculated by allocating proportional values to the original frequency categories with reference to a base value of 1.0, equivalent to once a day [20,41]. The scores were calculated as reported in Table 2.

Table 2. The six frequency response options and their conversion into Daily Equivalent Frequency.

Original Frequency used in FFQ	Daily Equivalent Frequency
Less than once per month	0.02
1–3 times per month	0.07
1–4 times per week	0.43
5–7 times per week	0.86
2–4 times per day	3.00
5 or more times per day	5.00

FFQ: Food Frequency Questionnaire.

Pearson's coefficients correlations were conducted to analyze the relationship between the gustatory functions.

Mixed ANOVAs were carried out considering "Taste sensitivity" (hyper, medium, and hypo) to basic tastes, "gender" (female, F and male; M) and their interaction as fixed factors and dietary intake, total energy and macronutrient intake as dependent variables, followed by pairwise comparisons using the Bonferroni test adjusted for multiple comparisons ($p < 0.05$). Participants were added as random factor in all the analyses. These statistical analyses were performed using IBM SPSS Statistics for Windows, Version 25.0 (IBM Corp., Armonk, NY, USA).

Correlation analyses between the tongue microbial ecology data, gustatory functions, and dietary intake, total energy and macronutrient intake were performed using the Kendall and Spearman formulas as predictors and dependent variables. Significance was set at $p \leq 0.05$ ($\alpha = 5\%$); significance in the range $0.05 < p < 0.10$ was accepted as a trend.

3. Results

3.1. Participant Characteristics

Fifty-nine volunteers (27 males, 32 females) were recruited for this study. The characteristics of all participants are detailed in Table 3.

Table 3. Baseline characteristics, energy and nutrient intake, and gustatory functions (taste thresholds, PROP responsiveness, and FPD) presented as mean, standard error of study participants.

	Mean	SEM
Age (years)	23.3	0.3
BMI (kg/m^2)	21.6	0.3
Gustatory Functions		
Sweet threshold (g/L)	3.61	0.62
Salty threshold (g/L)	0.20	0.02
Bitter threshold (g/L)	0.16	0.04
Sour threshold (g/L)	0.09	0.01
Food Intake		
Total Energy (kcal)	1829	60
Protein (%) [a]	15.6	0.4
Fat (%) [a]	35.2	0.9
Carbohydrates (%) [a]	45.2	0.7
Protein (g) [b]	68.0	2.3
Fat (g) [b]	70.1	2.7
Carbohydrates (g) [b]	216.0	8.6
Total Fiber (g) [b]	15.1	6.7

[a] Calculated as% of total energy intake (kcal); [b] Calculated as gram per day.

3.2. Association among Gustatory Functions and Their Relationship with Food Intake

Significant correlations were found between tastes that share many common features in the transduction mechanisms. In particular, the bitter threshold showed a significant correlation with the sweet threshold ($r = 0.34$, $p < 0.01$), while the sour threshold had significant correlations with the recognition thresholds of salty ($r = 0.34$, $p < 0.01$). Moreover, a significant correlation was found between sour and bitter thresholds ($r = 0.31$, $p < 0.05$).

As previously described in the "Material and Methods" section, basic taste thresholds were used as grouping variables and respondents were divided into three groups according to their sensitivity (hypersensitive, medium sensitive, hyposensitive). Among the whole samples, no more than 14 subjects switched from hypo- to hypersensitive within the different stimuli. For salty sensitivity, the group with low sensitivity corresponded to 27.1% of the total sample (10 M, 6 F), the medium sensitive group accounted for 27.1% (7 M, 9 F) and the group with high sensitivity corresponded to 45.8% (10 M, 17 F) of the total sample. For sweet sensitivity, the group with high sensitivity corresponded to 28.8% (9 M, 8 F) of the total sample, while the medium and hyposensitive groups accounted for 22.1% (6 M, 7 F) and 49.1% (12 M, 17 F) of the total sample, respectively. For bitter sensitivity, the hypersensitive group corresponded to 40.7% (15 M, 9 F) of the total sample, while the medium and hyposensitive groups accounted for 18.6% (4 M, 7 F) and 40.7% (8 M, 16 F) of the total sample, respectively. For sour sensitivity, the hypersensitive group corresponded to 22.0% (5 M, 8 F) of the total sample, while the medium and hyposensitive groups accounted for 44.1% (10 M, 17 F) and 33.9% (12 M, 8 F) of the total sample, respectively.

3.2.1. Salty Sensitivity

The elaboration of the results on potential impact of "*Salty sensitivity*" on food and beverage consumption frequency is reported in Table 4.

Consumption frequency of bakery and salty baked products, legumes, fats, and soft drinks seemed to be associated with "*Salty sensitivity*". Post hoc comparisons showed that, in general, hyposensitive subjects consumed these products significantly more than did medium and hypersensitive subjects. The main factor "*gender*" was significant for various food categories. In all cases, females have been found to consume significantly less salty baked products ($F_{(1,53)} = 8.46$, $p < 0.01$), cured meats ($F_{(1,53)} = 11.25$, $p < 0.001$), and soft drinks ($F_{(1,53)} = 10.19$, $p < 0.01$), but more fish ($F_{(1,53)} = 9.88$, $p < 0.01$), fruit ($F_{(1,53)} = 8.15$, $p < 0.01$), and nuts ($F_{(1,53)} = 7.02$, $p < 0.05$) than males. The "*Salty sensitivity*" × "*gender*" interaction was significant only in a few cases (cereal and cereal-derived products: $F_{(2,53)} = 5.52$, $p < 0.01$; cured meats: $F_{(2,53)} = 3.62$, $p < 0.05$; nuts: $F_{(2,53)} = 3.27$, $p < 0.05$; soft drinks: $F_{(2,53)} = 5.06$, $p < 0.01$).

When food record data were considered, a significant association with "*Salty sensitivity*" was found ($F_{(2,53)} = 3.52$, $p < 0.05$) on fat (as% energy intake), with hyposensitive subjects showing a higher intake compared to medium and hypersensitive subjects. A significant "*gender*" association was found ($F_{(2,53)} = 5.76$, $p < 0.05$), underlying a higher fat intake in female subjects with respect to males.

3.2.2. Sweet Sensitivity

The elaboration of the results on potential impact of "*Sweet sensitivity*" on food and beverage consumption frequency is reported in Table 4.

Consumption frequency of legumes and sweets and desserts seemed to be associated with "*Sweet sensitivity*". Post hoc comparisons showed that, in general, hypersensitive subjects consumed these products significantly less than did medium and hyposensitive subjects. The main factor "*gender*" was significant for various food categories. In all cases, females reported consuming significantly less salty baked products ($F_{(1,53)} = 14.29$, $p < 0.001$), cured meats ($F_{(1,53)} = 5.89$, $p < 0.05$), sweets and desserts ($F_{(1,53)} = 4.06$, $p < 0.05$), alcoholic beverages ($F_{(1,53)} = 5.19$, $p < 0.05$), and soft drinks ($F_{(1,53)} = 5.16$, $p < 0.05$), but more fish ($F_{(1,53)} = 5.70$, $p < 0.05$) than males. The "*Sweet sensitivity*" × "*gender*" interaction was significant only in a few cases (dairy products: $F_{(2,53)} = 3.48$, $p < 0.05$; candies and gums: $F_{(2,53)} = 3.17$, $p < 0.05$).

When food record data were considered, significant differences were found between female and male subjects on total energy and carbohydrates and fat consumptions. Energy (kcal) ($F_{(2,53)} = 4.71$, $p < 0.05$) and carbohydrate intakes (g) ($F_{(2,53)} = 5.70$, $p < 0.05$) were significantly lower in female subjects compared to male subjects. By contrast, fat intake (as% energy intake) ($F_{(2,53)} = 8.60$, $p < 0.01$) was significantly higher in females than males.

3.2.3. Bitter Sensitivity

The elaboration of the results on potential impact of "Bitter sensitivity" on food and beverage consumption frequency is reported in Table 5.

"*Bitter sensitivity*" had a significant association with consumption frequency of oils ($F_{(2,53)} = 5.41$, $p < 0.01$). Post hoc comparisons showed that hyposensitive subjects consumed these products significantly more than did medium and hypersensitive subjects. The main factor "*gender*" was significant for some food categories. In all cases, females reported consuming significantly less salty baked products ($F_{(1,53)} = 6.63$, $p < 0.05$) and cured meats ($F_{(1,53)} = 8.47$, $p < 0.01$), but more fish ($F_{(1,53)} = 8.79$, $p < 0.01$), fruit ($F_{(1,53)} = 4.87$, $p < 0.05$), and nuts ($F_{(1,53)} = 4.02$, $p < 0.05$) than males.

The "*Bitter sensitivity*" × "*gender*" interaction was significant only in a few cases (oils: $F_{(2,53)} = 5.32$, $p < 0.01$; salty snacks: $F_{(2,53)} = 3.54$, $p < 0.05$).

When food record data were considered, a significant association with "*Bitter sensitivity*" was found on energy (Kcal) ($F_{(2,53)} = 3.30$, $p < 0.05$) and carbohydrates (g) ($F_{(2,53)} = 3.59$, $p < 0.05$) intakes, with hypersensitive subjects showing higher intakes compared to medium and hyposensitive subjects. A significant "*gender*" association was found with carbohydrates ($F_{(2,53)} = 6.97$, $p < 0.01$) and fat intakes (as% energy intake) ($F_{(2,53)} = 11.77$, $p < 0.001$), underlying a lower carbohydrates but a higher fat intake in females than males.

3.2.4. Sour Sensitivity

The elaboration of the results on potential impact of "Sour sensitivity" on Food and Beverage consumption frequency is reported in Table 5.

"*Sour sensitivity*" had a significant association only with consumption frequency of fish ($F_{(2,53)} = 6.14$, $p < 0.01$). Post hoc comparisons showed that subjects characterized by medium sensitivity to sour consumed fish significantly less than did hypo- and hypersensitive subjects. The main factor "*gender*" was significant for some food categories. In all cases, females reported consuming significantly less salty baked products ($F_{(1,53)} = 8.21$, $p < 0.01$), cured meats ($F_{(1,53)} = 6.47$, $p < 0.05$), and soft drinks ($F_{(1,53)} = 5.29$, $p < 0.05$), but more fish ($F_{(1,53)} = 10.42$, $p < 0.01$) and nuts ($F_{(1,53)} = 4.04$, $p < 0.05$) than males. The "*Sour sensitivity*" × "*gender*" interaction was significant only in one category (salty snacks: $F_{(2,53)} = 3.54$, $p < 0.05$).

When food record data were considered, significant differences were found between female and male subjects for carbohydrates (g) ($F_{(2,53)} = 3.86$, $p < 0.05$) and fat consumptions (as% energy intake) ($F_{(2,53)} = 9.12$, $p < 0.01$). Carbohydrates intake (g) was significantly lower in female subjects compared to male subjects. By contrast, fat intake (as% energy intake) ($F_{(2,53)} = 8.60$, $p < 0.01$) was significantly higher in females than males.

3.3. *Correlation between Tongue Dorsum Microbiota, Gustatory Functions, and Dietary Intake*

To infer potential links between bacteria on tongue dorsum, gustatory functions, and dietary intake, we performed correlation analyses between taste thresholds, total energy, and macronutrient intake and the DADA2/SILVA/speciate11-determined bacterial relative abundances.

Several bacterial taxa abundances were correlated with taste thresholds. In particular, one taxon negatively correlated with sweet, three with sour, and six with salty thresholds. In summary, bacterial taxa abundances increase in subjects characterized by lower taste thresholds. On the contrary, the genus *Rothia* was the only taxon positively associated with taste thresholds, specifically salty. No taxon was correlated with the bitter threshold (Figure 1 – left side).

Finally, we performed correlation analyses between the oral microbiota and dietary intake. We found that energy and macronutrient intake were significantly correlated with several bacteria taxa (Figure 1 – right side). Notably, we found that several taxa in *Clostridia* class (e.g., genera *Selenomonas, Ruminococcaceae, Johnsonella*, and *Veilonella*) were positively correlated with total energy, protein, and fat intake and negatively correlated with carbohydrates and total fiber intake (e.g., genera *Catonella* and *Peptostreptococcus*). Contrarily, *Prevotella* genus was positively correlated with total fiber intake (Figure 1 – right side). Overall, these results seem to indicate that some microbial taxa are positively associated with vegetable-rich (*Prevotella* genus) or protein/fat-rich diets (*Clostridia* class).

Table 4. Mean values of daily equivalent frequency consumption for the 22 food categories by salty and sweet taste sensitivity level.

Items	*p value*	Daily Equivalent Frequency: Salty Taste Sensitivity Level			*p value*	Daily Equivalent Frequency: Sweet Taste Sensitivity Level		
		Hyper	Normal	Hypo		Hyper	Normal	Hypo
Cereal and cereal-derived products (e.g., pasta, rice, barley, spelt)	0.07	1.67	1.33	2.23	0.88	1.92	1.58	1.49
Salty baked products (e.g., bread, pizza, focaccia)	**0.007**	0.99 [b]	1.45 [ab]	1.99 [a]	0.69	1.37	1.12	1.40
Bakery products (e.g., bakery and breakfast cereals, biscuits, croissants)	**0.04**	0.76 [b]	1.39 [a]	1.42 [a]	0.48	1.12	1.34	0.94
Meats	0.98	0.50	0.51	0.49	0.85	0.48	0.54	0.50
Cured meats	0.64	0.36	0.40	0.31	0.09	0.27	0.48	0.37
Fish	0.16	0.41	0.28	0.43	0.35	0.39	0.28	0.39
Milk and yoghurts	0.47	0.84	0.75	1.11	0.27	0.75	0.61	1.04
Dairy products	0.32	0.51	0.79	0.79	0.19	0.52	0.62	0.86
Eggs	0.22	0.34	0.23	0.29	0.72	0.33	0.27	0.30
Vegetables	0.41	1.80	1.85	2.40	*0.08*	2.62	1.72	1.71
Legumes	**0.05**	0.36 [b]	0.67[a]	0.41 [ab]	**0.03**	0.69 [a]	0.41 [ab]	0.35 [b]
Potatoes	0.20	0.35	0.58	0.47	0.65	0.49	0.35	0.41
Fruit	0.22	1.65	1.84	2.42	0.79	2.07	1.78	1.78
Fruit juices	0.31	0.22	0.43	0.48	0.80	0.40	0.26	0.36
Nuts	0.27	0.23	0.45	0.36	0.76	0.33	0.22	0.33
Sweets and desserts (e.g., cakes, ice creams, chocolate)	0.53	0.60	0.86	0.78	**0.02**	0.32 [b]	0.98 [a]	0.87 [a]
Fats	**0.05**	0.26 [b]	0.22 [b]	0.45 [a]	0.18	0.19	0.35	0.34
Oils	0.47	1.99	1.54	1.72	0.87	1.83	1.94	1.74
Salty snacks (e.g., chips, salty peanuts)	0.23	0.22	0.22	0.35	0.23	0.17	0.27	0.30
Alcoholic beverages	0.21	0.49	0.29	0.31	0.18	0.30	0.56	0.35
Soft drinks	**0.007**	0.26 [b]	1.13 [a]	0.16 [b]	0.82	0.59	0.39	0.38
Candies and gums	0.82	0.63	0.54	0.78	0.32	0.82	0.81	0.40

Significant *p*-values are shown in bold. Different letters indicate significant differences according to Bonferroni's post hoc test.

Table 5. Mean values of daily equivalent frequency consumption for the 22 food categories by bitter and sour taste sensitivity level.

Items	Daily Equivalent Frequency				Daily Equivalent Frequency			
	p value	Bitter Taste Sensitivity Level			*p value*	Sour Taste Sensitivity Level		
		Hyper	Normal	Hypo		Hyper	Normal	Hypo
Cereal and cereal-derived products (e.g., pasta, rice, barley, spelt)	0.58	1.75	1.29	1.66	0.08	2.19	1.31	1.78
Salty baked products (e.g., bread, pizza, focaccia)	0.70	1.45	1.25	1.20	0.11	0.90	1.43	1.66
Bakery products (e.g., bakery and breakfast cereals, biscuits, croissants)	0.81	0.92	1.10	1.10	0.21	0.69	1.20	1.29
Meats	0.65	0.47	0.48	0.54	0.75	0.53	0.51	0.46
Cured meats	0.41	0.34	0.46	0.33	0.37	0.35	0.40	0.29
Fish	0.22	0.42	0.26	0.36	**0.004**	0.49 [a]	0.27 [b]	0.46 [a]
Milk and yoghurt	0.66	0.72	0.97	0.92	0.96	0.91	0.84	0.82
Dairy products	0.07	0.52	0.52	0.95	0.55	0.49	0.75	0.69
Eggs	0.52	0.33	0.24	0.30	0.54	0.36	0.31	0.27
Vegetables	0.08	1.92	1.04	2.23	0.33	1.84	1.77	2.39
Legumes	0.51	0.53	0.42	0.38	0.09	0.32	0.61	0.40
Potatoes	0.53	0.49	0.39	0.35	0.23	0.61	0.42	0.36
Fruit	0.76	2.06	1.70	1.81	0.94	1.77	1.94	1.94
Fruit juices	0.24	0.48	0.39	0.19	0.70	0.22	0.31	0.39
Nuts	0.42	0.43	0.26	0.26	0.41	0.43	0.23	0.36
Sweets and desserts (e.g., cakes, ice creams, chocolate)	0.85	0.73	0.61	0.78	0.44	0.46	0.75	0.79
Fats	0.84	0.29	0.29	0.25	0.26	0.26	0.36	0.22
Oils	**0.007**	1.46 [b]	1.11 [b]	2.22 [a]	0.35	1.64	1.72	2.16
Salty snacks (e.g., chips, salty peanuts)	0.40	0.24	0.20	0.32	0.44	0.18	0.28	0.22
Alcoholic beverages	0.34	0.46	0.33	0.29	0.70	0.31	0.43	0.37
Soft drinks	0.22	0.68	0.20	0.16	0.81	0.40	0.56	0.36
Candies and gums	0.91	0.68	0.73	0.57	0.76	0.48	0.63	0.78

Significant *p*-values are shown in bold. Different letters indicate significant differences according to Bonferroni's post hoc test.

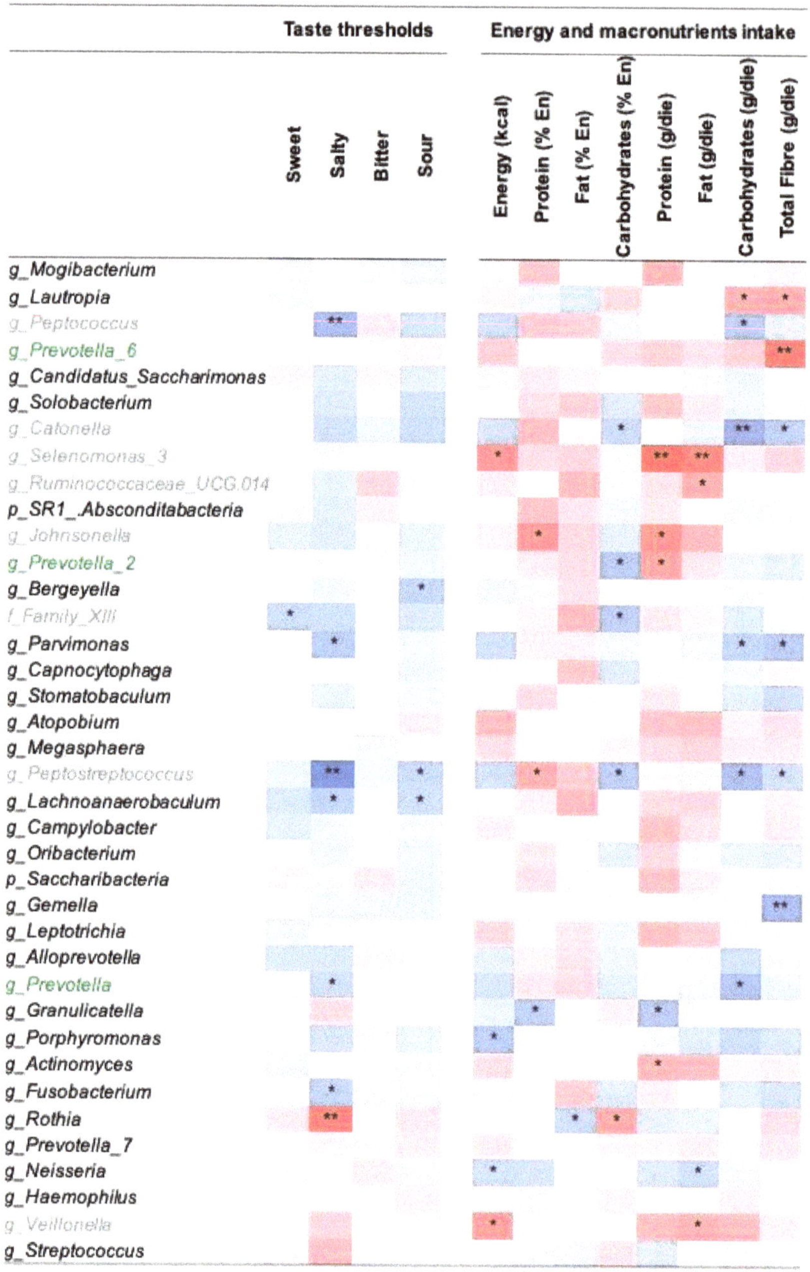

Figure 1. Correlations between the relative abundance of bacterial taxa on tongue dorsum and taste thresholds for the four basic tastes (left side) and nutritional variables (right side). The heatmap represents the Spearman's correlation R-values. Asterisks relate to the Kendall rank correlation p values: * $p < 0.05$; ** $p < 0.01$. *Prevotella* genera and *Clostridia* class, which showed significant correlation with energy and macronutrients intake, are highlighted in green and in grey colors, respectively.

4. Discussion

The present study evaluated interindividual differences in recognition thresholds for basic tastes and examined to what extent these variations in gustatory functions among individuals could be related to food intake in a sample of reportedly healthy adult women and men. Additionally, these variables were further evaluated in relation to individual oral microbiota composition.

The present study shows that recognition thresholds for the basic tastes were associated with each other, albeit in different ways. Indeed, significant correlations were found between tastes that share many common features in the transduction mechanisms. The perception of both sweet and bitter tastes is mediated via G-coupled protein receptors, encoded by TAS1R and TAS2R taste receptor gene families, while salty and sour tastes are transduced via ion channels [42]. Thus, these findings seem to confirm the presence of the well-known dichotomy in taste coding for perception of pleasant (e.g., sweet and savory compounds) vs noxious stimuli (e.g., sour and bitter tastants) [43].

The results of the present study showed that interindividual differences in taste perception may influence habitual food consumption and intake. This assumption supports various observations that taste sensitivity may play an important role in dietary habits and body energy balance [37,44–46]. Indeed, it has been suggested that subjects characterized by a reduced or distorted taste sensitivity could increase the willingness to ingest foods that involve greater stimulation of the taste and oral somatosensory system (e.g., high-energy dense foods rich in sugars and fats), leading to unhealthy food choices, and thus pathogenesis of weight excess.

As far as salt intake is regarded, a few studies [47–49] examined the association between salt hedonics and sodium intake, but the relationship between salty sensitivity and food intake has been poorly investigated. Kim and Lee [25] reported an association between individuals' salty taste sensitivity and sodium-rich fast foods acceptance and consumption in a sample of Korean adolescents. Moreover, Hayes and colleagues [47] reported that variation in salt perception was associated with differences in preferences to high-sodium foods and, indirectly, to sodium intake. Accordingly, in the present study, subjects who were orally hypersensitive to sodium chloride solution reported consuming less bakery and salty baked products than those who were defined as hyposensitive. Moreover, hyposensitivity to salty taste seems to increase consumption of less healthy foods, like saturated-fat-rich products and soft drinks. This assumption is supported by food record data, in which fat intake (expressed as a percentage of total energy intake) was found to be higher in the hyposensitive group.

Previous studies have failed to find associations between sweet taste and diet parameters [21,23]. Contrarily, the present data from the FB-FQ and sweet sensitivity suggests that participants who have a higher threshold for sweet taste (hyposensitive) reported consuming more frequently sweets and desserts than the hypersensitive group. This is supported by findings of recent studies, where positive relationships were found between reduced perceived intensity and increased desire for higher energy providing taste stimuli [50]. Accordingly, Jayasinghe and colleagues [20] showed that participants who perceived as sweeter the highest glucose concentrations are reported to be more sensitive to sweet taste and had a lower consumption frequency of sweet foods compared to those who perceived the solution proposed as less sweet.

Interestingly, this study did not find any relationship between sweet taste sensitivity and energy or macronutrient intakes expressed as a percentage of total energy or grams. This result was in contrast with previous findings suggesting an inverse correlation between glucose taste perception and total energy and carbohydrate intakes [20]. However, as recently discussed by Webb and colleagues [38] it is necessary to use a combination of sweet taste measurements (e.g., glucose, sucrose, and fructose) to better characterize the overall perception and the relationships between sweet taste perception and food intake.

Nevertheless, we observed a relationship between bitter taste sensitivity and total energy and carbohydrates intakes. Participants who were orally hypersensitive to caffeine solutions showed higher total energy and carbohydrate intakes compared to those who perceived the solution as less bitter, suggesting a potential shift towards less healthy dietary patterns in the hypersensitive group of subjects. These results seem to support the hypothesis that higher taste sensitivity to bitter compounds can elicit rejection responses in subjects leading to a reduced selection and intake of some vegetable foods in favour of high-energy-dense foods [51].

Regarding the relationship between sour taste sensitivity and food consumption frequency, our results failed to underline any significant association. It is important to note that, in literature, the attention has been mainly focused on individual variation in sour taste perception and preferences for sour foods [52], suggesting that low preference for sour foods could eventually lead to limited choices or inadequate intake of fruit and berries. However, even if these results demonstrated a genetic contribution to preference for sour foods, the authors underlined that sour taste perception and related preferences for sour foods are mediated by both genetic and environmental factors (e.g., food habits of the family). Thus, the potential relationship among sour taste perception and subsequent food choices and intake remains to be explored [16].

As expected, gender-related differences in food consumption frequency and intake were found, confirming previous studies in which differences in the nutritional quality of the diet of men and women were highlighted [53–56]. Indeed, men reported significantly more frequent consumption of salty baked products, cured meats, sweets and desserts, alcoholic beverages, and soft drinks than did women. On the other hand, women were more likely to consume fish, fruit and nuts. Macronutrients intake, in terms of percentage of total energy intake, differed between female and male subjects. Clearly, as expected, men consumed higher total energy and carbohydrates compared to women.

In order to provide further insights into the complexities of human eating behavior, the present study focused the attention on less investigated nongenetic factors potentially influencing food preference and habits. In particular, we considered the oral microbiota composition since, recently, a relationship between reduced taste perception and specific oral bacteria's growth has been reported [35], in agreement with previous findings [33,34]. Solemdal and colleagues [34] investigated variables related to taste ability and oral health in acutely hospitalized elderly, showing that taste perception, particularly for sour taste, was reduced in acutely hospitalized elderly with high lactobacilli growth. Besnard and colleagues [33] tested the hypothesis that obese and normal-weight adults could be characterized by an impaired fat taste perception, which could be also linked to a change in the microbial composition. This study showed no difference in the fat taste perception and composition of oral microbiota between normal-weight and obese subjects. Otherwise, specific bacterial composition was found in lipid non-tasters, irrespectively of nutritional status. Moreover, in our previous study, we found that subjects who were characterized by a greater responsiveness to PROP presented differences in the relative abundance of some taxa compared to subjects who were less responsive to the PROP compound. In particular, five bacterial genera, including the Gram-positive genera *Actinomyces*, *Oribacterium*, *Solobacterium*, and *Catonella*, and the Gram-negative *Campylobacter*, were overrepresented in the most responsiveness group.

In the present study, interesting further correlations between the relative abundance of bacterial taxa on tongue dorsum and gustatory functions were found. The present results showed that a number of taxa were inversely correlated with salt and sour thresholds, showing that a great salty and sour sensitivity may be linked to specific taxa, mainly attributed to *Clostridiales* and *Bacteroidales* order. Given the diversity of genera and species within the oral microbiome [57], it is overall difficult to propose systematic explanations of such links. However, a hypothesis may reside in bacterial modulatory ability as suggested by Alcock and colleagues [58], who described a potential involvement of microbes in the manipulation of eating behavior by altering the host preferences through a modulation of receptor expression, as in vivo animal model studies on gut microbiota showed [59,60]. Another plausible explanation may lie in bacterial ability to degrade carbohydrates into disaccharides, monosaccharides, and organic acids, used as "building material" for biofilms [61]. The physical barrier between tastants and taste receptors would, as a consequence, be less or more efficient, thus influencing sensitivity. According to our results, also Feng and colleagues [62] reported that a higher proportion of *Actinobacteria* was linked to lower taste sensitivity, while a higher proportion of *Bacteroidetes* increased sensitivity.

Nonetheless, a more detailed characterization of microbial communities and their metabolic feature would be of interest, but this study supports that the oral microbiota composition deserves to be considered as an influencing variable when investigating peri-receptor events involved in chemosensory processes.

The role of diet in shaping the gut microbiota is widely recognized [63,64]. However, until recently, only a few studies have considered the association between habitual diet and oral microbiota.

In the present study, interesting correlations between the relative abundance of bacterial taxa on tongue dorsum and dietary intake were found. Indeed, *Clostridia* class was positively associated with total energy, fat, and protein intake but negatively associated with fiber intake, whereas *Proteobacteria* phylum and *Prevotella* genus showed the opposite association. Since it has been found that oral cavity and stool bacteria overlapped in nearly half (45%) of the subjects in recent studies [65,66], it is possible to hypothesize that dietary habits could affect both oral and gut microbiota in a similar way. Indeed, our results are in line with the general assumption that some gut microbial taxa are positively associated with vegetable-rich (*Prevotella*) or protein/fat-rich diets (*Clostridia*) [67,68]. However, further studies are warranted to clarify whether observations from the gut microbiome are transferrable to the oral microbiome. In this context, the oral microbiome could be further investigated as potential marker of long-term consumption of healthy or unhealthy diets.

The strengths of the present study include an investigation of the relationship between taste sensitivity for all the four tastes with a range of parameters of food consumption frequency and food intake. In particular, dietary intake was investigated through assessment of actual food intakes (seven-day food records) and habitual intakes of different categories of foods and beverages (FB-FQ), capturing different aspects of eating habits. Finally, the multidisciplinary approach applied in the present study offers new insights into the reciprocal impact between taste perception, food intake, and oral microbiota composition.

The present study has also several limitations. Firstly, participants involved were a small sample of Italian women and men of similar age (young) and BMI (normal range). Therefore, the findings of this study cannot be generalized to other ethnicities, ages, or BMI groups. Secondly, the study design was cross-sectional and the findings represent only relationships among variables under study while no causations can be ascertained. Thirdly, limitations to the study include validity of food intake measurements. Reported intakes may be inaccurate due to memory recall, interviewer and subject bias, and responder fatigue, all of which contribute to underestimating or overestimating food intake measures [69].

In conclusion, the present study shows a link between taste sensitivity and dietary measurements in a group of young healthy women and men with normal BMI and food intake. Moreover, significant relationships between taste sensitivity and dietary measurements, but also with oral microbiota composition, were found.

These findings have implications for eating behavior, as perceived sensory properties of foods and beverages clearly influence preferences and the type and amount of food consumed [1]. Moreover, this study provides further support that nongenetic factors, such as the oral bacteria lining the tongue, should be adequately considered in order to gain new insights into taste-related eating habits that may influence long-term health outcomes. The impact of genetic and nongenetic characteristics, including the complex interactions among multiple factors related with food cues and exposure, can affect food choices and dietary intake. For this reason, this topic remains an important research area to be further investigated, since all these aspects reciprocally influence each other, driving towards individual eating behavior.

Author Contributions: Conceptualization, C.C., P.R., M.L., and E.P.; Formal analysis, C.C., P.R., and G.G.; Main Funding acquisition, M.L.; Investigation, C.C.; Project administration, P.R. and E.P.; Supervision, E.P.; Visualization, G.G.; Writing—original draft, C.C.; Writing—review & editing, P.R., M.L., G.G., and E.P. All authors read and approved the final manuscript.

Funding: This study was funded by the University of Milan project "Sviluppo Unimi" - Linea B: Istituzione di fondi di ricerca da assegnare a giovani ricercatori (Project title: UNderstandIng Childhood Obesity (UNICORN): Perceptive and behavioral determinants of weight gain in children; Project n. PSR2017-MLAUR granted to M.L.) and it was supported by 'Rose Marie Pangborn Sensory Science Scholarship Fund (SSSF, granted to C.C.) and the contribution from MAPLE project (Mipaaft grant D.M. 8245/7303/2016 granted to P.R.) within the European Joint Programming Initiative "A Healthy Diet for a Healthy Life" (JPI- HDHL - http://www.healthydietforhealthylife.eu/).

Acknowledgments: We thank Gabriela Mazzitelli for her help in the analysis of food diaries and all the volunteers.

Conflicts of Interest: The author declares no conflict of interest.

References

1. Mennella, J.A.; Pepino, M.Y.; Reed, D.R. Genetic and environmental determinants of bitter perception and sweet preferences. *Pediatrics* **2005**, *115*, e216–e222. [CrossRef]
2. Chaudhari, N.; Roper, S.D. The cell biology of taste. *J. Cell Biol.* **2010**, *190*, 285–296. [CrossRef]
3. Reed, D.R.; Tanaka, T.; McDaniel, A.H. Diverse tastes: Genetics of sweet and bitter perception. *Physiol Behav.* **2006**, *88*, 215–226. [CrossRef]
4. Van Dongen, M.V.; van den Berg, M.C.; Vink, N.; Kok, F.J.; de Graaf, C. Taste–nutrient relationships in commonly consumed foods. *Br. J. Nutr.* **2012**, *108*, 140–147. [CrossRef]
5. Tepper, B.J. Nutritional implications of genetic taste variation: The role of PROP sensitivity and other taste phenotypes. *Annu. Rev. Nutr.* **2008**, *28*, 367–388. [CrossRef]
6. Bartoshuk, L.M. The biological basis of food perception and acceptance. *Food Qual. Prefer.* **1993**, *4*, 21–32. [CrossRef]
7. Dinnella, C.; Monteleone, E.; Piochi, M.; Spinelli, S.; Prescott, J.; Pierguidi, L.; Gasperi, F.; Laureati, M.; Pagliarini, E.; Predieri, S. Individual variation in PROP status, fungiform papillae density, and responsiveness to taste stimuli in a large population sample. *Chem. Senses.* **2018**, *43*, 697–710. [CrossRef]
8. Bajec, M.R.; Pickering, G.J. Thermal taste, PROP responsiveness, and perception of oral sensations. *Physiol. Behav.* **2008**, *95*, 581–590. [CrossRef]
9. Bere, E.; Brug, J.; Klepp, K.I. Why do boys eat less fruit and vegetables than girls? *Public Health Nutr.* **2008**, *11*, 321–325. [CrossRef]
10. Feeney, E.; O'brien, S.; Scannell, A.; Markey, A.; Gibney, E.R. Genetic variation in taste perception: Does it have a role in healthy eating? *Proc. Nutr. Soc.* **2011**, *70*, 135–143. [CrossRef]
11. Rasmussen, M.; Krølner, R.; Klepp, K.I.; Lytle, L.; Brug, J.; Bere, E.L.; Due, P. Determinants of fruit and vegetable consumption among children and adolescents: A review of the literature. Part I: Quantitative studies. *Int. J. Behav. Nutr. Phys. Act.* **2006**, *3*, 22. [CrossRef]
12. Dinehart, M.E.; Hayes, J.E.; Bartoshuk, L.M.; Lanier, S.L.; Duffy, V.B. Bitter taste markers explain variability in vegetable sweetness, bitterness, and intake. *Physiol. Behav.* **2006**, *87*, 304–313. [CrossRef]
13. Drewnowski, A.; Henderson, S.A.; Shore, A.B.; Barratt-Fornell, A. Sensory responses to 6-n-propylthiouracil (PROP) or sucrose solutions and food preferences in young women. *Ann. N. Y. Acad. Sci.* **1998**, *855*, 797–801. [CrossRef]
14. Jerzsa-Latta, M.; Krondl, M.; Coleman, P. Use and perceived attributes of cruciferous vegetables in terms of genetically-mediated taste sensitivity. *Appetite* **1990**, *15*, 127–134. [CrossRef]
15. Yackinous, C.A.; Guinard, J.X. Relation between PROP (6-n-propylthiouracil) taster status, taste anatomy and dietary intake measures for young men and women. *Appetite* **2002**, *38*, 201–209. [CrossRef]
16. Garcia-Bailo, B.; Toguri, C.; Eny, K.M.; El-Sohemy, A. Genetic variation in taste and its influence on food selection. *OMICS* **2009**, *13*, 69–80. [CrossRef]
17. Tan, S.Y.; Tucker, R. Sweet Taste as a predictor of dietary intake: A systematic review. *Nutrients* **2019**, *11*, 94. [CrossRef]
18. Cicerale, S.; Riddell, L.J.; Keast, R.S. The association between perceived sweetness intensity and dietary intake in young adults. *J. Food Sci.* **2012**, *77*, H31–H35. [CrossRef]
19. Holt, S.H.A.; Cobiac, L.; Beaumont-Smith, N.E.; Easton, K.; Best, D.J. Dietary habits and the perception and liking of sweetness among Australian and Malaysian students: A cross-cultural study. *Food Qual. Prefer.* **2000**, *11*, 299–312. [CrossRef]
20. Jayasinghe, S.N.; Kruger, R.; Walsh, D.C.; Cao, G.; Rivers, S.; Richter, M.; Bernhard, H.B. Is sweet taste perception associated with sweet food liking and intake? *Nutrients* **2017**, *9*, 750. [CrossRef]

21. Low, J.Y.; Lacy, K.E.; McBride, R.; Keast, R.S. The association between sweet taste function, anthropometry, and dietary intake in adults. *Nutrients* **2016**, *8*, 241. [CrossRef]
22. Mahar, A.; Duizer, L.M. The effect of frequency of consumption of artificial sweeteners on sweetness liking by women. *J. Food Sci.* **2007**, *72*, S714–S718. [CrossRef]
23. Martinez-Cordero, E.; Malacara-Hernandez, J.M.; Martinez-Cordero, C. Taste perception in normal and overweight Mexican adults. *Appetite* **2015**, *89*, 192–195. [CrossRef]
24. Durack, E.; Alonso-Gomez, M.; Wilkinson, M.G. Salt: A review of its role in food science and public health. *Curr. Nutr. Food Sci.* **2008**, *4*, 290–297. [CrossRef]
25. Kim, G.H.; Lee, H.M. Frequent consumption of certain fast foods may be associated with an enhanced preference for salt taste. *J. Hum. Nutr. Diet.* **2009**, *22*, 475–480. [CrossRef]
26. Kobayashi, C.; Kennedy, L.M.; Halpern, B.P. Experience-induced changes in taste identification of monosodium glutamate (MSG) are reversible. *Chem. Senses.* **2006**, *31*, 301–306. [CrossRef]
27. Pittman, D.W.; Contreras, R.J. Dietary NaCl influences the organization of chorda tympani neurons projecting to the nucleus of the solitary tract in rats. *Chem. Senses.* **2002**, *27*, 333–341. [CrossRef]
28. Wise, P.M.; Hansen, J.L.; Reed, D.R.; Breslin, P.A. Twin study of the heritability of recognition thresholds for sour and salty taste. *Chem. Senses* **2007**, *32*, 749–754. [CrossRef]
29. Running, C.A.; Craig, B.A.; Mattes, R.D. Oleogustus: The unique taste of fat. *Chem. Senses* **2015**, *40*, 507–516. [CrossRef]
30. Tucker, R.M.; Kaiser, K.A.; Parman, M.A.; George, B.J.; Allison, D.B.; Mattes, R.D. Comparisons of fatty acid taste detection thresholds in people who are lean vs. overweight or obese: A systematic review and meta-analysis. *PLoS ONE* **2017**, *12*, e0169583. [CrossRef]
31. Keast, R.S.; Costanzo, A. Is fat the sixth taste primary? Evidence and implications. *Flavour* **2015**, *4*, 5. [CrossRef]
32. Martínez-Ruiz, N.R.; López-Díaz, J.A.; Wall-Medrano, A.; Jiménez-Castro, J.A.; Angulo, O. Oral fat perception is related with body mass index, preference and consumption of high-fat foods. *Physiol. Behav.* **2014**, *129*, 36–42. [CrossRef]
33. Besnard, P.; Christensen, J.E.; Brignot, H.; Bernard, A.; Passilly-Degrace, P.; Nicklaus, S.; Pais de Barros, J.P.; Collet, X.; Lelouvier, B.; Servant, F. Obese subjects with specific gustatory papillae microbiota and salivary cues display an impairment to sense lipids. *Sci. Rep.* **2018**, 8. [CrossRef]
34. Solemdal, K.; Sandvik, L.; Willumsen, T.; Mowe, M.; Hummel, T. The impact of oral health on taste ability in acutely hospitalized elderly. *PLoS ONE* **2012**, *7*, e36557. [CrossRef]
35. Cattaneo, C.; Gargari, G.; Koirala, R.; Laureati, M.; Riso, P.; Guglielmetti, S.; Pagliarini, E. New insights into the relationship between taste perception and oral microbiota composition. *Sci Rep.* **2019**, 9. [CrossRef]
36. Hardikar, S.; Höchenberger, R.; Villringer, A.; Ohla, K. Higher sensitivity to sweet and salty taste in obese compared to lean individuals. *Appetite* **2017**, *111*, 158–165. [CrossRef]
37. Proserpio, C.; Laureati, M.; Bertoli, S.; Battezzati, A.; Pagliarini, E. Determinants of obesity in Italian adults: The role of taste sensitivity, food liking, and food neophobia. *Chem. Senses.* **2016**, *41*, 169–176. [CrossRef]
38. Webb, J.; Bolhuis, D.P.; Cicerale, S.; Hayes, J.E.; Keast, R. The relationships between common measurements of taste function. *Chemosens Percept.* **2015**, *8*, 11–18. [CrossRef]
39. International Organisation for Standardization. Sensory analysis—Methodology—General guidance for measuring odour, flavour and taste detection thresholds by a three-alternative forced-choice (3-AFC) procedure. In *ISO International Standard*; N° 13301/International Organization for Standardizatio: Geneva, Switzerland, 2018; p. 28.
40. Porrini, M.; Gentile, M.G.; Fidanza, F. Biochemical validation of a self-administered semi-quantitative food-frequency questionnaire. *Br. J. Nutr.* **1995**, *74*, 323–333. [CrossRef]
41. Daly, A.M.; Parsons, J.E.; Wood, N.A.; Gill, T.K.; Taylor, A.W. Food consumption habits in two states of Australia, as measured by a Food Frequency Questionnaire. *BMC Res. Notes.* **2011**, *4*, 507. [CrossRef]
42. Drayna, D. Human taste genetics. *Annu. Rev. Genomics Hum. Genet.* **2005**, *6*, 217–235. [CrossRef]
43. Hladik, C.M.; Pasquet, P.; Simmen, B. New perspectives on taste and primate evolution: The dichotomy in gustatory coding for perception of beneficent versus noxious substances as supported by correlations among human thresholds. *Am. J. Phys. Anthropol.* **2002**, *117*, 342–348. [CrossRef]

44. Bertoli, S.; Laureati, M.; Battezzati, A.; Bergamaschi, V.; Cereda, E.; Spadafranca, A.; Laila, V.; Ella, P. Taste sensitivity, nutritional status and metabolic syndrome: Implication in weight loss dietary interventions. *World J. Diabetes.* **2014**, *5*, 717. [CrossRef]
45. Donaldson, L.F.; Bennett, L.; Baic, S.; Melichar, J.K. Taste and weight: Is there a link? *Am. J. Clin. Nutr.* **2009**, *90*, S800–S803. [CrossRef]
46. Proserpio, C.; Laureati, M.; Invitti, C.; Pagliarini, E. Reduced taste responsiveness and increased food neophobia characterize obese adults. *Food Qual. Prefer.* **2018**, *63*, 73–79. [CrossRef]
47. Hayes, J.E.; Sullivan, B.S.; Duffy, V.B. Explaining variability in sodium intake through oral sensory phenotype, salt sensation and liking. *Physiol. Behav.* **2010**, *100*, 369–380. [CrossRef]
48. Mattes, R.D. The taste for salt in humans. *Am. J. Clin. Nutr.* **1997**, *65*, 692S–697S. [CrossRef]
49. Pangborn, R.M.; Pecore, S.D. Taste perception of sodium chloride in relation to dietary intake of salt. *Am. J. Clin. Nutr.* **1982**, *35*, 510–520. [CrossRef]
50. Noel, C.A.; Sugrue, M.; Dando, R. Participants with pharmacologically impaired taste function seek out more intense, higher calorie stimuli. *Appetite* **2017**, *117*, 74–81. [CrossRef]
51. Stevenson, R.J.; Boakes, R.A.; Oaten, M.J.; Yeomans, M.R.; Mahmut, M.; Francis, H.M. Chemosensory abilities in consumers of a western-style diet. *Chem. Senses.* **2016**, *41*, 505–513. [CrossRef]
52. Törnwall, O.; Silventoinen, K.; Keskitalo-Vuokko, K.; Perola, M.; Kaprio, J.; Tuorila, H. Genetic contribution to sour taste preference. *Appetite* **2012**, *58*, 687–694. [CrossRef]
53. Fagerli, R.A.; Wandel, M. Gender differences in opinions and practices with regard to a 'healthy diet'. *Appetite* **1999**, *32*, 171–190. [CrossRef]
54. Leblanc, V.; Bégin, C.; Corneau, L.; Dodin, S.; Lemieux, S. Gender differences in dietary intakes: What is the contribution of motivational variables? *J. Hum. Nutr. Diet.* **2015**, *28*, 37–46. [CrossRef]
55. Li, K.K.; Concepcion, R.Y.; Lee, H.; Cardinal, B.J.; Ebbeck, V.; Woekel, E.; Readdy, R.T. An examination of sex differences in relation to the eating habits and nutrient intakes of university students. *J. Nutr. Educ. Behav.* **2012**, *44*, 246–250. [CrossRef]
56. Rolls, B.J.; Fedoroff, I.C.; Guthrie, J.F. Gender differences in eating behavior and body weight regulation. *Health Psychol.* **1991**, *10*, 133–142. [CrossRef]
57. Dewhirst, F.E.; Chen, T.; Izard, J.; Paster, B.J.; Tanner, A.C.; Yu, W.H.; Lakshmanan, A.; Wade, W.G. The human oral microbiome. *J. Bacteriol.* **2010**, *192*, 5002–5017. [CrossRef]
58. Alcock, J.; Maley, C.C.; Aktipis, C.A. Is eating behavior manipulated by the gastrointestinal microbiota? Evolutionary pressures and potential mechanisms. *Bioessays* **2014**, *36*, 940–949. [CrossRef]
59. Duca, F.A.; Swartz, T.D.; Sakar, Y.; Covasa, M. Increased oral detection, but decreased intestinal signaling for fats in mice lacking gut microbiota. *PLoS ONE* **2012**, *7*, e39748. [CrossRef]
60. Swartz, T.D.; Duca, F.A.; De Wouters, T.; Sakar, Y.; Covasa, M. Up-regulation of intestinal type 1 taste receptor 3 and sodium glucose luminal transporter-1 expression and increased sucrose intake in mice lacking gut microbiota. *Br. J. Nutr.* **2012**, *107*, 621–630. [CrossRef]
61. Takahashi, N. Oral microbiome metabolism: From "who are they?" to "what are they doing?". *J. Dent. Res.* **2015**, *94*, 1628–1637. [CrossRef]
62. Feng, Y.; Licandro, H.; Martin, C.; Septier, C.; Zhao, M.; Neyraud, E.; Martine, M. The associations between biochemical and microbiological variables and taste differ in whole saliva and in the film lining the tongue. *BioMed. Res. Int.* **2018**. [CrossRef]
63. Jeffery, I.B.; O'Toole, P.W. Diet-microbiota interactions and their implications for healthy living. *Nutrients* **2013**, *5*, 234–252. [CrossRef]
64. Albenberg, L.G.; Wu, G.D. Diet and the intestinal microbiome: Associations, functions, and implications for health and disease. *Gastroenterol* **2014**, *146*, 1564–1572. [CrossRef]
65. Olsen, I.; Yamazaki, K. Can oral bacteria affect the microbiome of the gut? *J. Oral Microbiol.* **2019**, *11*, 1586422. [CrossRef]
66. Segata, N.; Haake, S.K.; Mannon, P.; Lemon, K.P.; Waldron, L.; Gevers, D.; Huttenhower, C.; Izard, J. Composition of the adult digestive tract bacterial microbiome based on seven mouth surfaces, tonsils, throat and stool samples. *Genome. Biol.* **2012**, *13*, R42. [CrossRef]
67. Wu, G.D.; Chen, J.; Hoffmann, C.; Bittinger, K.; Chen, Y.Y.; Keilbaugh, S.A.; Bewtra, M.; Knights, D.; Walters, W.A.; Knight, R. Linking long-term dietary patterns with gut microbial enterotypes. *Science* **2011**, *334*, 105–108. [CrossRef]

68. Turnbaugh, P.J.; Ridaura, V.K.; Faith, J.J.; Rey, F.E.; Knight, R.; Gordon, J.I. The effect of diet on the human gut microbiome: A metagenomic analysis in humanized gnotobiotic mice. *Sci. Transl. Med.* **2009**, *1*, 6ra14. [CrossRef]
69. Kretsch, M.J.; Fong, A.K.; Green, M.W. Behavioral and body size correlates of energy intake underreporting by obese and normal-weight women. *J. Am. Diet. Assoc.* **1999**, *99*, 300–306. [CrossRef]

nutrients

MDPI

Article

Sodium Imbalance in Mice Results Primarily in Compensatory Gene Regulatory Responses in Kidney and Colon, but Not in Taste Tissue

Kristina Lossow [1,2], Wolfgang Meyerhof [1,3] and Maik Behrens [1,4,*]

1 Molecular Genetics, German Institute of Human Nutrition Potsdam-Rehbruecke, Arthur-Scheunert-Allee 114-116, 14558 Nuthetal, Germany; kristina.lossow@uni-jena.de (K.L.); wolfgang.meyerhof@uks.eu (W.M.)
2 Department of Molecular Nutritional Physiology, Institute of Nutritional Sciences, Friedrich Schiller University Jena, Dornburger Str. 24, 07743 Jena, Germany
3 Center for Integrative Physiology and Molecular Medicine, Saarland University, Kirrbergerstr., Bldg. 48, 66421 Homburg, Germany
4 Leibniz-Institute for Food Systems Biology at the Technical University of Munich, Lise-Meitner-Str. 34, 85354 Freising, Germany
* Correspondence: m.behrens.leibniz-lsb@tum.de

Received: 12 March 2020; Accepted: 30 March 2020; Published: 3 April 2020

check for updates

Abstract: Renal excretion and sodium appetite provide the basis for sodium homeostasis. In both the kidney and tongue, the epithelial sodium channel (ENaC) is involved in sodium uptake and sensing. The diuretic drug amiloride is known to block ENaC, producing a mild natriuresis. However, amiloride is further reported to induce salt appetite in rodents after prolonged exposure as well as bitter taste impressions in humans. To examine how dietary sodium content and amiloride impact on sodium appetite, mice were subjected to dietary salt and amiloride intervention and subsequently analyzed for ENaC expression and taste reactivity. We observed substantial changes of ENaC expression in the colon and kidney confirming the role of these tissues for sodium homeostasis, whereas effects on lingual ENaC expression and taste preferences were negligible. In comparison, prolonged exposure to amiloride-containing drinking water affected β- and αENaC expression in fungiform and posterior taste papillae, respectively, next to changes in salt taste. However, amiloride did not only change salt taste sensation but also perception of sucrose, glutamate, and citric acid, which might be explained by the fact that amiloride itself activates bitter taste receptors in mice. Accordingly, exposure to amiloride generally affects taste impression and should be evaluated with care.

Keywords: epithelial sodium channel; sodium homeostasis; amiloride; salt deprivation; short-term preference test

1. Introduction

Sodium is the main cation in the extracellular fluid and the primary determinant of osmolarity, crucial for many biological processes [1]. Accordingly, levels of sodium are tightly controlled through a precise balance of sodium intake and excretion. The latter is primarily realized via the kidney, which plays a major role in volume, electrolyte, and blood pressure regulation [2,3]. To move sodium across the apical plasma membrane, the distal tubules and collecting duct of the kidney utilize the epithelial sodium channel (ENaC). In times of sodium deficits, aldosterone, a mineralocorticoid hormone produced in the adrenal glands [4], promotes the translocation of ENaC from cytoplasmic compartments to the apical plasma membrane in the renal collecting system [5] triggering sodium reabsorption. In comparison, sodium absorption in the intestine is primarily realized by the sodium/hydrogen exchanger rather than through the ENaC, which is limited to epithelial cells of the distal colon and rectum [6,7]. However, after proctocolectomy, ENaC was verified in the distal part of the small intestine, indicating the importance of ENaC-mediated reabsorption of salt and water in the intestine [8]. Inhibition of ENaC is achieved by amiloride [9–12], first described by Cragoe in 1967 [13]. Amiloride lowers systemic blood pressure by preventing absorption of sodium and increasing its excretion along with water (natriuresis) [14–16].

Additionally, sodium homeostasis is maintained by ingestive behavior. Sodium appetite is the instinctive drive to seek salty substances or beverages for consumption [17] stimulated by sodium deficiency, hypovolemia, or mineralocorticoids. Accordingly, sodium-depleted rats ingest high sodium chloride solutions even at concentrations they would normally reject [18–21]. Already decades ago, studies revealed that amiloride not only affects the kidney but also acts as a potent blocker of salt taste in rodents [22–25]. Applied to the tongue before or during sodium stimulation, amiloride reduces sodium-evoked responses in the chorda tympani nerve innervating the fungiform papillae on the frontal tongue as well as the palate in the oral cavity [9,10,26,27]; vice versa, sodium deprivation or aldosterone application results in an increased number of amiloride-sensitive taste receptor cells in fungiform papillae [28].

Conclusively, ENaC has been detected in lingual epithelia and taste buds [9,10,23,29–33]. Fully functional ENaC is believed to be composed of three homologous subunits (α, β, and γ) arranged with a 1:1:1 stoichiometry [34,35]. In 2010, Chandrashekar and colleagues demonstrated the involvement of the α-subunit of ENaC in salt attraction and sodium taste responses by a tissue-specific knock-out model [32]. However, a recent gene-targeted approach, using fluorescent marker proteins under the control of the *Scnn1a* and *Scnn1b* gene loci, encoding α- and βENaC, respectively, concluded that the assumed $\alpha\beta\gamma$-subunit composition of ENaC seems highly unlikely in taste tissue, as ENaC subunits were distributed in different taste bud cells under adequate sodium conditions [33].

Accordingly, the cellular and molecular composition of the "salt taste receptor" is still quite controversial with many unanswered questions. Furthermore, studies regarding the expression levels of ENaC subunits in gustatory tissues after dietary sodium restriction are rare and inconclusive [28,36]. To explore how sodium ingestion as well as short- and longtime exposure to amiloride impact on sodium appetite, we utilized a gene-targeted animal model with modified *Scnn1a* and *Scnn1b* loci, which were subjected to dietary and amiloride interventions, followed by taste reactivity testing (short-term preference test) and expression analysis of potential ENaC subunits.

2. Materials and Methods

2.1. Animal Experiments

All animal experiments were approved by and conducted following the national guidelines of the Ministry of Environment, Health and Consumer Protection of the federal state of Brandenburg (14513 Teltow, Germany; 2347-02-2017), and institutional guidelines of the German Institute of Human Nutrition Potsdam-Rehbruecke (14558 Nuthetal, Germany; T-01-17, T-02-17). Mice were housed in polycarbonate cages and kept under constant conditions (12 h light/dark cycles with light beginning at 6:00 am; 22 °C room temperature; 55% humidity). The animals received food and water *ad libitum* (for details see feeding experiments).

2.2. Feeding Experiment and Amiloride Intervention

In this study, either wild-type or homozygous double gene-targeted Scnn1a$^{IRES\text{-}GFP}$/Scnn1b$^{IRES\text{-}tdRFP}$ mice (abbreviated Scnn1$^{++/++}$ and Scnn1$^{aa/bb}$, respectively, throughout) with modified *Scnn1a* and *Scnn1b* loci were used. These carried modified alleles allowing the synthesis of green (GFP) and red (tdRFP) fluorescent proteins in cells expressing α- and βENaC subunits, respectively. Details of modification/gene-targeting were described previously [33].

Before the experiments, Scnn1$^{++/++}$ and Scnn1$^{aa/bb}$ animals had free access to water and standard chow diet (V1534, Ssniff, Soest, Germany; sodium content of 0.24%). From the 7th week of life on, the animals continued to have free access to water, while the chow was replaced either by a sodium-adequate (equal sodium content as within the chow, however 4.8 above nutrient requirements of laboratory animals) [37], sodium-deficient or high salt diet, with 0.21% (0.5% NaCl; E15430-04, Ssniff, Soest, Germany), <0.03% (E15430-24, Ssniff, Soest, Germany), and 1.71% sodium (4% NaCl; E15431-34, Ssniff, Soest, Germany), respectively. Body weight, food, and water intake were monitored weekly from weaning onward (21 days). Blood pressure was recorded prior to diet change/experimental session (6th week) and after 4 weeks of intervention (10th week) using Power Lab and Chart 5 software (ADInstruments Ltd., Oxford, UK). Urine samples were collected during the day (6 am to 5 pm) and overnight (5 pm to 6 am) at week 10, while mice were individually housed in metabolic cages (TechniPlast, Buguggiate, Italy). Urinary sodium content was determined in triplicate with a LAQUA twin system (Horiba Scientific for Na+, Darmstadt, Germany). Until then, all animals went through the same experimental procedure differing only in the sodium content of the diet. To test the effect of the consumed sodium on RNA expression levels and the distribution of fluorescent proteins, a randomly selected experimental set of mice were killed at the 11th week of life and subjected either to taste bud isolation (n = 4–6 per diet and genotype) or perfusion (n = 3 per diet and genotype). In a different experimental set, taste responses to different taste solutions were assessed by a short-term preference test. The remaining mice further continued with the dietary intervention for additional 4 weeks (n = 10–11 per diet and genotype) with defined access to water and food (for details see short-term preference test). In the third experimental setup, taste responses to different taste solutions after pre-treatment with amiloride were examined. In this experiment, mice fed a sodium-adequate diet either received access to amiloride-containing (300 μM) or amiloride-free water for 36 h [38] prior to short-term preference tests (n = 10–16 per intervention and genotype). Amiloride treatment was restricted to a maximum of 3 times for each animal, with a recovery phase of 36 h in-between (with amiloride-free water). After the final intervention/third amiloride treatment, animals were subjected either to taste bud isolation (n = 4 per intervention and genotype) or perfusion (n = 3 per intervention and genotype).

2.3. Short-term Preference Tests

Short-term preference tests were performed using the DavisRig system (MS-160, DiLog Instruments, Tallahassee, FL, USA), permitting several taste stimuli to be presented in a brief trial within a single test session. During these experiments, mice were singly housed with defined access to water and food. Mice were initially trained for 3 days after 18 h water restriction, using water as test stimulus to get used to the shutter system. To monitor attractive taste responses, animals were restricted for 22.5 h with limited access to food (1 g of the corresponding diet) and water (2.5 mL). For aversive taste stimuli, mice were water restricted over a period of 22.5 h with free access to food, as reported earlier [39]. Mice performed preference tests at the beginning of the light phase, followed by 1 h with free access to water and food. Test sessions were restricted to 2 or 5 consecutive days for attractive or aversive taste stimuli, respectively. To control for small individual differences in lick rate, number of licks to each taste stimulus was divided by the average number of licks to water alone for each mouse and day. A lick ratio of 1.0 reveals that licks to taste stimulus equals licks to water. In contrast, a lick ratio close to 0 indicates only a few licks of taste stimulus relative to licks for water, and a lick ratio > 1.0 signifies more licks of the taste stimulus relative to that for water. In case of apparent decreasing motivation of the animals, for example, reduced water licks under aversive test conditions or diminished licks for sucrose and sucralose under attractive restriction conditions, we excluded the entire subsequent test session. Accordingly, sometimes only the first 15 instead of 20 min were taken into consideration, adjusted day by day, and according to the motivational behavior of each mouse.

Taste solutions for testing were prepared daily with reagent grade chemicals and distilled water, sucrose (10–1000 mM, Merck, Darmstadt, Germany), monopotassium glutamate (MPG; 1–100 mM; Fluka, Oberhaching, Germany), denatonium benzoate (DB; 0.1–10 mM; Sigma, Taufkirchen, Germany), citric acid (CA; 1–100 mM; Roth, Karlsruhe, Germany), sodium chloride (NaCl; 10–1000 mM;

Roth, Karlsruhe, Germany), inosine monophosphate (IMP; 1 mM; Sigma, Taufkirchen, Germany), and amiloride (0.1 mM; Sigma, Taufkirchen, Germany). Solutions were presented at room temperature. The sequence of stimulus presentation was randomized with varying taste solutions and concentrations (no ascending or descending presentation of a single taste solution) every day to minimize systematic order and contrast effects.

2.4. Tissue Preparation

$Scnn1^{++/++}$ and $Scnn1^{aa/bb}$ mice were either anesthetized by intraperitoneal injection of 150 mg/kg body weight pentobarbital (Narcoren from Merial, Hallbergmoos, Germany) followed by transcardial perfusion with phosphate-buffered saline (PBS; 0.01 M Na_2HPO_4, 1.764 mM KH_2PO_4, 2.683 mM KCl, 0.1369 M NaCl, pH 7.4) and 4% paraformaldehyde (PFA) to gain tissue for cryosections (14 μm) and subsequent immunohistochemistry or with isoflurane (Cp-pharma, Burgsdorf, Germany) to gain tissues for RNA extractions as reported earlier [33].

2.5. Immunohistochemistry

Immunohistochemistry was performed as described before [33]. Primary antibodies included TrpM5 (1:5000; [40]) and aromatic L-amino acid decarboxylase (AADC, also known as DOPA decarboxylase, 1:500; GTX30448, GeneTex, Irvine, CA, USA).

Co-localization analysis in fungiform papillae was based on confocal (Leica TCS SP8) z-stack scans through tissue sections, of which a single plane was used for evaluation. Mean fluorescence intensity was evaluated in the operating mode of TCS SP8 LAS X software by defining a region of interest (ROI), namely one taste bud, in the digitized graph. Collected data reflect results from tissue sections of at least 3 mice per intervention.

2.6. RNA Isolation and qRT-PCR

Taste cells, lingual epithelium, kidney, and distal colon were subjected to RNA isolation and cDNA synthesis as previously reported [33]. Afterwards, 12.5 ng/well cDNA served as a template for quantitative RT-PCR using TaqMan Gene Expression Master Mix (Applied Biosystems by ThermoFisher Scientific, Foster City, CA, USA) and the corresponding probes (αENaC, Mm00803386_m1; βENaC, Mm00441215_m1; γENaC, Mm00441228_m1; Eef2, Mm01171435_gH; all ABI Applied Biosystems by ThermoFisher Scientific, Foster City, CA, USA; β-actin probe 5′FAM-TTGAGACCTTCAACACCCCAGCCA-3′TAM, β-actin for 5′-TACGACCAGAGGCATACAG-3′, β-actin rev 5′-GCCAACCGTGAAAAGATGAC-3′; Eurofins MWG Operon, Martinsried, Germany) in a final volume of 10 μL. After initial 10 min denaturation at 95 °C, PCR was carried out for 40 cycles with 95 °C for 15 s and 60 °C for 60 s using Quant Studio 12K Flex Real-Time PCR System (Applied Biosystems by ThermoFisher Scientific, Foster City, CA, USA). Relative expression was determined based on normalization to β-actin and eukaryotic translation elongation factor 2 (Eef2) mRNA levels. As amplification efficiencies of all used probes were quite similar, cycle threshold (CT) values were averaged from triplicates and differences between CT values of ENaC and the housekeeping genes were calculated as ΔCT for normalization and finally expressed as $2^{-\Delta CT}$. Collected data reflect results from isolated tissue of at least 4 mice.

2.7. Functional Expression Analysis

Functional expression analysis of bitter taste receptor cDNAs has been reported in detail earlier [39]. For the screening for amiloride-responsive bitter taste receptors, the compound has been applied to human embryonal kidney cells transiently expressing mouse (Tas2rs) and human bitter taste receptors (TAS2Rs) at a concentration of 3 mM. Subsequently, hits were verified by dose-concentration analyses, with concentrations ranging between 0.0001 and 3 mM. Latter data reflect results from 3 individual experiments.

2.8. Statistics

Data are given as means ± standard deviation (STABW; tables) or standard error (STE; figures). Statistical reliability of the observed results in the feeding study was determined by UNIANOVA and post-hoc analysis using Bonferroni's multiple comparison test (SPSS, SPSS 16.0, IBM, Chigago, IL, USA). For quantitative expression analysis, significant differences between 3 diets were determined by UNIANOVA (SPSS, SPSS 16.0, IBM, Chicago, IL, USA), followed by a comparison of individual pairs of means using Bonferroni's post-hoc test, whereas comparison between 2 intervention groups was conducted by using Student's unpaired *t*-test. Differences were considered to be statistically significant if $p \leq 0.05$.

3. Results

3.1. Feeding Experiment

To investigate if diets with varying sodium content impact on the expression of ENaC, we fed Scnn1$^{++/++}$ and Scnn1$^{aa/bb}$ mice with low, adequate, and high salt diets over a period of 4 weeks (from 7th to 10th week of age). We monitored body weight, food, and water intake from weaning onward. As expected, body weight constantly increased during the experiment independent of diet and genotype ($p < 0.001$; Figure 1A). Mice consumed equal amounts of food, ranging from 1.3 g after weaning to 3.7 g, with a mean intake of 2.5 g per day, irrespective of the sodium content of the diet (Figure 1B). In contrast, water intake changed after the diet switched (Figure 1C). When fed a high salt diet, mice increased their daily water intake by about 78% in comparison to animals fed with a diet adequate in sodium. The water intake of mice on low, adequate, and high salt diet ranged around 3.6 ± 0.5 mL, 3.6 ± 0.6 mL, and 6.5 ± 0.7 mL per day, respectively ($p < 0.001$). Furthermore, we observed significant differences in water intake based on genotype from the 5th week on, with significantly higher values in Scnn1$^{aa/bb}$ animals (Figure 1C).

In addition, we measured blood pressure prior and at the end of the dietary intervention (Figure 1D). Prior to diet change, no significant differences for mean systolic blood pressure were observed. After 3.5 weeks, animals that received the high salt diet showed slightly increased systolic blood pressure compared to animals that ate low or adequate salt diets, triggering significant diet X genotype effects ($p = 0.012$). Furthermore, urine was collected in metabolic cages at the end of the dietary intervention. Urinary volumes of mice fed the high salt diet were slightly higher, relative to the other groups. However, differences did not reach significance when considered for day, night, or total volume, due to high variations between the animals (Figure 1E). Considering mean urinary sodium content, significant variances were observed depending on diet (significant differences between all groups) and diet X genotype ($p < 0.001$), but not based on genotype itself. Thereby, no significant differences between groups fed low and sodium-adequate diet were recognized, whereas animals receiving the high salt diet showed significantly higher urinary sodium content in comparison to all other groups (Figure 1F).

3.2. Expression Analysis

In order to examine the impact of dietary salt on the expression of the ENaC subunits, we measured the transcription levels in the gustatory tissue, kidney, and distal colon of Scnn1$^{aa/bb}$ mice. Quantitative RT-PCR analyses of isolated fungiform and foliate/vallate taste buds and of non-gustatory lingual tissue showed no differences in relative expression levels of all 3 ENaC subunits after feeding a high or low salt diet in comparison to Scnn1$^{aa/bb}$ animals fed a control diet over a period of 4 weeks (Table 1). In kidney, αENaC expression was significantly reduced after consumption of a high salt diet, whereas β- and γENaC were not affected (Table 1). In the distal colon, intervention with a low salt diet resulted in significantly higher expression levels of α- and βENaC in comparison to Scnn1$^{aa/bb}$ animals fed with a standard or high salt diet. For γENaC, such significant variances were only detected between

Scnn1$^{aa/bb}$ animals fed a low and high salt diet, with higher expression levels for animals who received a low salt diet (Table 1).

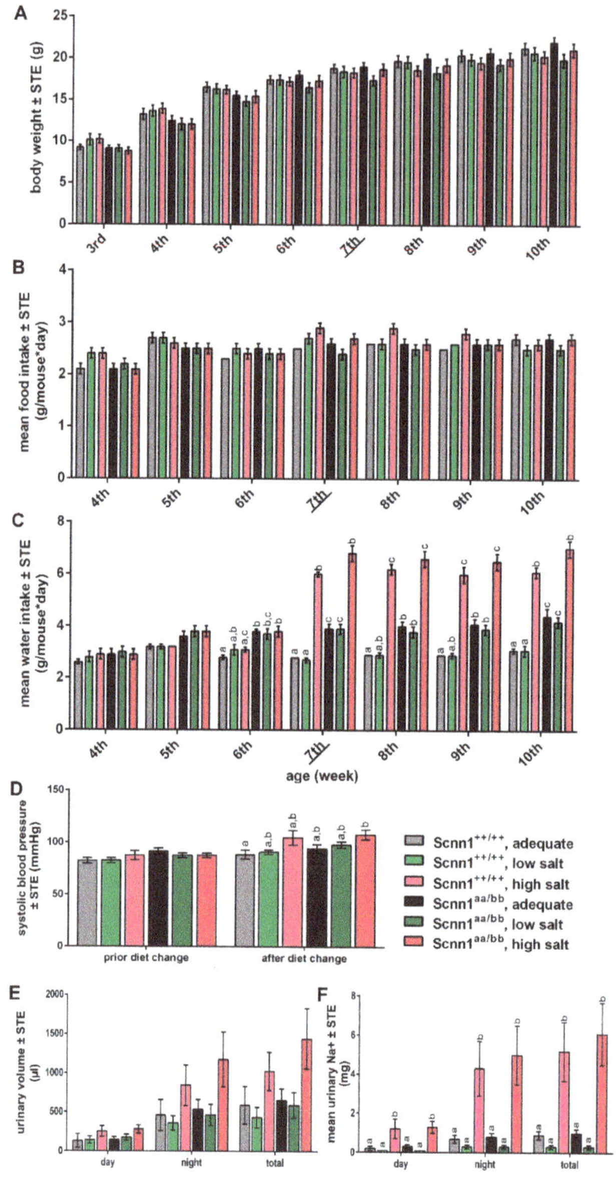

Figure 1. Physiological parameters of Scnn1$^{++/++}$ and Scnn1$^{aa/bb}$ mice during dietary monitoring. (**A**) Body weight, (**B**) food and (**C**) water intake, (**D**) blood pressure, (**E**) urinary volume, and (**F**) sodium excretion of Scnn1$^{++/++}$ (n = 11) and Scnn1$^{aa/bb}$ mice (n = 19–20) fed with low salt, adequate, and high salt diet over a period of 4 weeks (7th to 10th week, time point of diet change is indicated by underline); after initial 3 weeks of sodium adequate chow diet (4th to 6th week) after weening (3rd week). Statistical differences between 2 bars at a specific time point are indicated by different letters based on UNIANOVA and post-hoc analysis using Bonferroni´s multiple comparison test.

Table 1. Relative expression of epithelial sodium channel (ENaC) subunits in Scnn1[aa/bb] mice after dietary intervention. Based on quantitative RT-PCR, the relative expression levels of ENaC subunits normalized to the housekeeping genes β-actin and eEf2 in isolated taste buds and non-gustatory tissue were determined. Data represent the means of 6 Scnn1[aa/bb] mice, fed with sodium-adequate, low, or high salt diet over a period of 4 weeks. Statistical testing is based on UNIANOVA and post-hoc analysis using Bonferroni´s multiple comparison test. *p*-Values rely on comparison of all groups with significant differences indicated in bold; individual differences between the different diets are additionally indicated as the following: * statistical significance between sodium-adequate and high salt diet; # statistical significance between sodium-adequate and low salt diet; $ statistical significance between low and high salt diet based on Student´s *t*-test.

	ENaC Subunit	Adequate (mean ± STABW)	Low Salt (mean ± STABW)	High Salt (mean ± STABW)	*p*-Value
fungiform papillae	α	0.0151 ± 0.0053	0.0174 ± 0.0098	0.0201 ± 0.0104	0.735
	β	0.0062 ± 0.0030	0.0101 ± 0.0059	0.0087 ± 0.0069	0.464
	γ	0.0093 ± 0.0026	0.0118 ± 0.0066	0.0117 ± 0.0075	0.719
vallate and foliate papillae	α	0.0606 ± 0.0111	0.0577 ± 0.0217	0.0538 ± 0.0152	0.780
	β	0.0027 ± 0.0007	0.0024 ± 0.0010	0.0036 ± 0.0009	0.084
	γ	0.0020 ± 0.0005	0.0018 ± 0.0007	0.0021 ± 0.0004	0.669
non-gustatory epithelium	α	0.0228 ± 0.0061	0.0283 ± 0.0107	0.0205 ± 0.0037	0.212
	β	0.0011 ± 0.0004	0.0010 ± 0.0003	0.0010 ± 0.0002	0.942
	γ	0.0008 ± 0.0001	0.0007 ± 0.0003	0.0007 ± 0.0002	0.891
kidney	α	0.0698 ± 0.0235	0.0600 ± 0.0154	0.0420 ± 0.0069	**0.034** *
	β	0.0260 ± 0.0086	0.0196 ± 0.0043	0.0215 ± 0.0043	0.211
	γ	0.0425 ± 0.0112	0.0296 ± 0.0036	0.0360 ± 0.0084	0.054
distal colon	α	0.0357 ± 0.0104	0.0504 ± 0.0102	0.0266 ± 0.0063	**0.002** #,$
	β	0.0051 ± 0.0017	0.0258 ± 0.0184	0.0026 ± 0.0010	**0.003** #,$
	γ	0.0091 ± 0.0068	0.0306 ± 0.0302	0.0027 ± 0.0026	**0.041** $

Furthermore, no differences were noted regarding expression of the fluorescent reporter proteins in fungiform taste buds of Scnn1[aa/bb] animals. Consumption of either the low or high salt diet did not affect the location of GFP and tdRFP fluorescence, nor the fluorescence intensity. Accordingly, GFP always co-localized with Type III cell marker AADC, whereas tdRFP was not co-expressed with Type II or Type III cell markers TrpM5 or AADC, respectively (Figure 2).

However, ENaC expression of Scnn1[aa/bb] in comparison to Scnn1[++/++] animals consuming an adequate salt diet revealed overall higher expression levels (Table S1). Whereas non-gustatory tissue differences in the expression levels of ENaC subunits did not reach statistical significance, prominent differences were recognized in lingual taste papillae affecting all ENaC subunits. In kidney, genotype primarily affected α- and βENaC subunits, whereas in the distal colon only βENaC expression was significantly increased in Scnn1[aa/bb] animals.

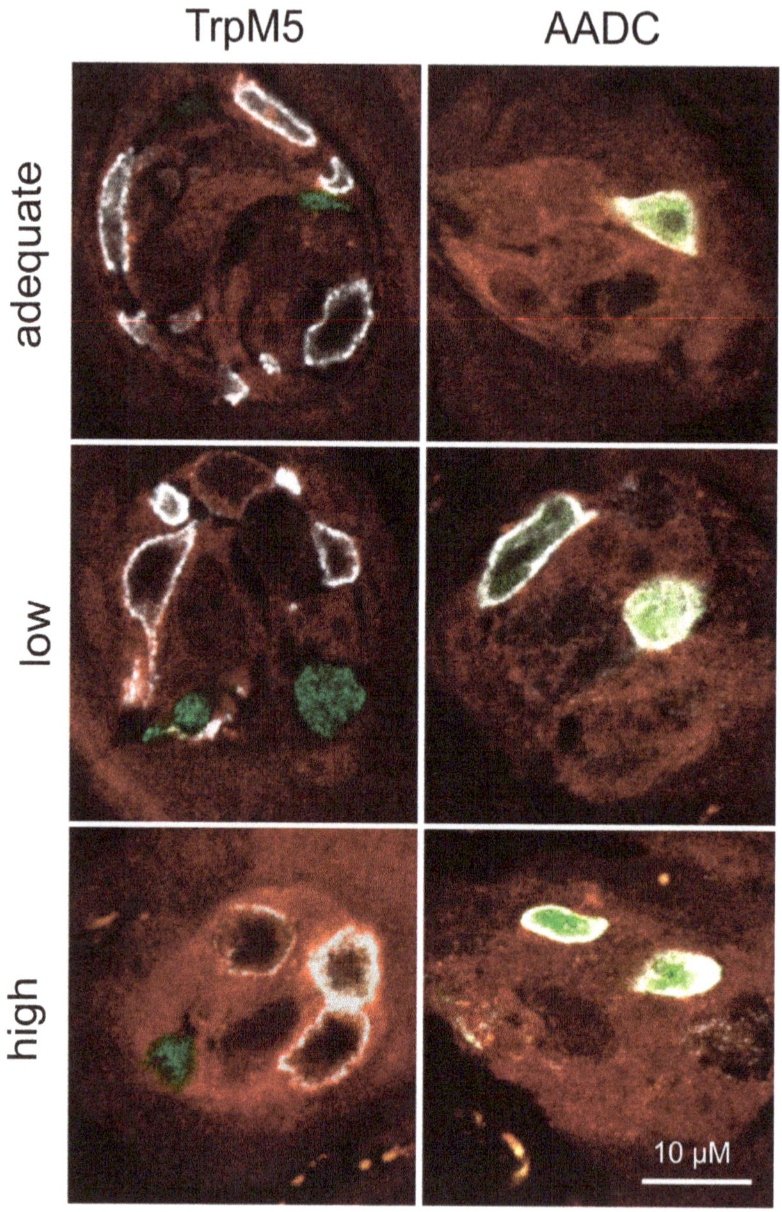

Figure 2. Expression of fluorescent proteins in fungiform papillae after dietary intervention. Fungiform papillae sections of Scnn1$^{aa/bb}$ animals expressing GFP (synthesis of green) and tdRFP (synthesis of red) fluorescence in αENaC- and βENaC-expressing cells, respectively, were stained for Type II (TrpM5) and Type III (AADC) taste cell markers after dietary intervention. Therefore, animals received either an adequate, low, or high salt diet over a period of 4 weeks. Independently of the consumed diet, GFP and tdRFP fluorescence showed no co-localization in taste papillae. Whereas GFP-positive cells always co-expressed AADC, tdRFP-positive cells revealed no overlap with the cell markers TrpM5 or AADC, visualized by immunofluorescence (white). Scale bars apply to all images.

3.3. Short-term Preference Tests

In order to determine if the sodium content of diets affects taste preferences for NaCl or other tastants, despite the lack of changes in gustatory ENaC expression, we performed a dietary intervention study. A group of 10 to 11 Scnn1$^{++/++}$ and Scnn1$^{aa/bb}$ mice was fed an adequate, low, or high salt diet for another 4 weeks (total intervention of 8 weeks, from week 7–15). During this time, short-term preference tests were performed under conditions allowing the analysis of taste responses to attractive (Figure 3A–D) and aversive (Figure 3E–H) taste stimuli. For this procedure, it was mandatory to deprive animals of water. Even though the mice that received the high salt diet drank significantly more water under *ad libitum* conditions (Figure 1C), water restriction did not have an impact on the water lick rates during 5 s measuring periods. Accordingly, taste preference tests were carried out under the same conditions for all animals, independent of diet or genotype. Mean lick responses are shown in Figure 3.

Under conditions for testing attractive stimuli there was no diet X genotype effect for sucrose (Table S3). However, ANOVA revealed an effect for concentration with increasing number of licks/water at 30 mM sucrose for Scnn1$^{aa/bb}$ animals under the high salt regime (Figure 3A). For MPG+IMP, there were effects for concentration and diet X genotype (Figure 3B, Tables S2 and S3). The latter was observed at 1.0, 3.0, and 10 mM, with Scnn1$^{aa/bb}$ animals showing increased lick rates (Table S2). NaCl taste solutions were licked more at the low concentration of 30 mM by Scnn1$^{aa/bb}$ animals under the high salt regime and resulted in higher lick numbers at the high salt concentration of 300 mM by Scnn1$^{aa/bb}$ animals pretreated according to the low salt regime (Figure 3C, Tables S2 and S3). These effects were eliminated by amiloride treatment (Figure 3D, Tables S2 and S3). Control stimuli like IMP and denatonium benzoate presented only at a single concentration did not result in any diet X genotype effects (Figure S1A, Tables S2 and S3). In comparison to that, the control stimuli amiloride and citric acid revealed such effects, with higher lick/water ratios in Scnn1$^{aa/bb}$ animals (Figure S1A, Tables S2 and S3). For amiloride, highest lick/water means were observed for Scnn1$^{aa/bb}$ animals fed with a high salt diet, which were even higher than for animals fed the low or adequate diet, indicating that in this case not only genotype but also diet has an effect.

Additionally, animals were further water-restricted for 22.5 h to perform short-term preference tests for aversive stimuli (Figure 3E–H, Tables S2 and S3). Here, all taste stimuli were affected by concentration and diet X genotype (Table S3). For denatonium benzoate, Scnn1$^{++/++}$ mice showed higher mean licks for denatonium benzoate at 0.1 to 1.0 mM than other diet X genotype constellations (Figure 3E). Scnn1$^{++/++}$ animals further tended to show higher lick to water ratios for citric acid at 1 and 10 mM (exception: low salt fed animals at 1 mM; Figure 3F). In comparison to that, NaCl did not result in significant differences at 10 to 100 mM, whereas Scnn1$^{aa/bb}$ and Scnn1$^{++/++}$ mice fed with a low salt diet and Scnn1$^{aa/bb}$ animals fed with an adequate diet showed higher lick ratios than Scnn1$^{++/++}$ animals fed with an adequate diet or animals fed with a high salt diet (both genotypes; Figure 3G). Independent of diet, we observed changes in lick ratios among Scnn1$^{++/++}$ and Scnn1$^{aa/bb}$ animals for 1 mM MPG+IMP, 100–1000 mM NaCl (attractive regime) as well as for 10 mM and 100 mM citric acid, 300–1000 mM NaCl, and 1000 mM NaCl+amiloride (aversive regime), respectively, when testing diet adequate in sodium content ($p \leq 0.05$, Student's *t*-test).

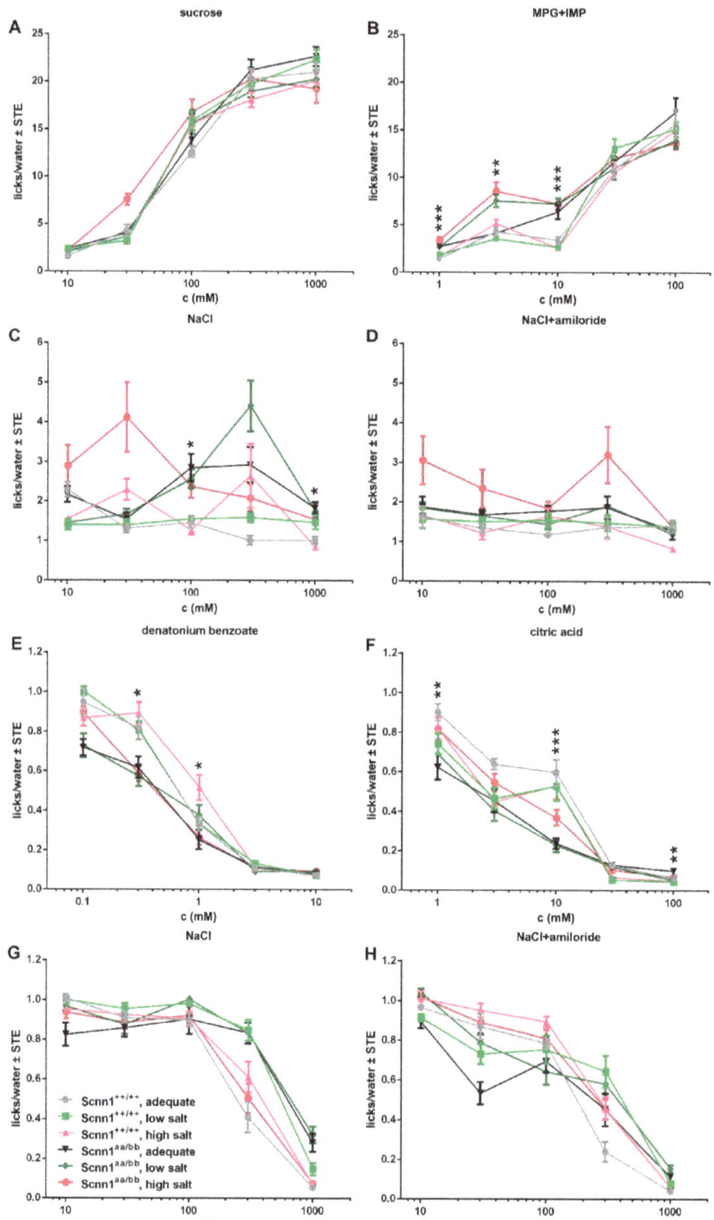

Figure 3. Taste response curves of Scnn1$^{++/++}$ and Scnn1$^{aa/bb}$ mice after dietary intervention. After 4 weeks fed with sodium-adequate, low, or high salt diet, Scnn1$^{++/++}$ and Scnn1$^{aa/bb}$ mice were subjected to short-term preference tests using an automated gustometer. To do so, animals were either restricted for 22.5 h with access to 2.0 mL water and 1 g food (attractive restriction conditions, (**A–D**)) or water-deprived for 22.5 h (aversive restriction conditions, (**E–H**)). Taste solutions and concentrations were presented randomly. Each data point represents a mean ± standard error (SE) of 5 s presentations from the 10 to 11 animals tested. Statistical testing was based on UNIANOVA and post-hoc analysis using Bonferroni´s multiple comparison test. Significant differences over all groups in line drawings were indicated by asterisk(s) with * $p < 0.05$, ** $p < 0.01$, and *** $p < 0.001$.

3.4. Amiloride Intervention

As reported in a recent study, application of 300 μM amiloride over a period of 36 h induces robust salt appetite [38]. To investigate whether amiloride affects only the preference for sodium or also ENaC expression, Scnn1$^{++/++}$ and Scnn1$^{aa/bb}$ mice were subjected to short-term preference tests as well as to taste bud and tissue preparations after receiving amiloride for 36 h next to a sodium-adjusted diet. According to the latter and subsequent expression analysis, free access to amiloride-containing water did not have an impact on the relative expression values for α-, β- or γENaC subunits in non-gustatory tissue and kidney, whereas in fungiform and foliate/vallate papillae, β- and αENaC, respectively, were induced after amiloride treatment (Table 2). However, no impact of amiloride intervention on GFP and tdRFP fluorescence was recognized in fungiform papillae with regard to localization and apparent intensity (Figure 4). According to expression analysis, all 3 ENaC subunits of the distal colon were affected by amiloride intervention, with significantly higher relative expression levels after access to amiloride for 36 h (Table 2).

Table 2. Relative expression of ENaC subunits in Scnn1$^{aa/bb}$ mice after amiloride intervention. Based on quantitative RT-PCR the relative expression levels of ENaC subunits normalized to the housekeeping genes β-actin and eEf2 in isolated taste buds and non-gustatory tissue were determined. Data represent means of 6 Scnn1$^{aa/bb}$ mice, with access to a sodium-adequate diet and water or 300 μM amiloride-containing drinking solution for 36 h prior to tissue isolation. Statistical testing was based on Student´s *t*-test. Differences were considered to be significant if $p < 0.05$, as indicated in bold.

	ENaC Subunit	Water (mean ± STABW)	Amiloride (mean ± STABW)	*p*-Value
fungiform papillae	α	0.0104 ± 0.0019	0.0136 ± 0.0019	0.051
	β	0.0030 ± 0.0014	0.0060 ± 0.0012	**0.017**
	γ	0.0056 ± 0.0025	0.0063 ± 0.0024	0.700
vallate and foliate papillae	α	0.0529 ± 0.0089	0.0700 ± 0.0083	**0.031**
	β	0.0024 ± 0.0003	0.0028 ± 0.0006	0.311
	γ	0.0018 ± 0.0004	0.0017 ± 0.0003	0.853
non-gustatory epithelium	α	0.0229 ± 0.0041	0.0280 ± 0.0019	0.066
	β	0.0008 ± 0.0004	0.0013 ± 0.0004	0.195
	γ	0.0007 ± 0.0002	0.0008 ± 0.0003	0.682
kidney	α	0.0583 ± 0.0217	0.0744 ± 0.0199	0.317
	β	0.0223 ± 0.0061	0.0238 ± 0.0057	0.738
	γ	0.0399 ± 0.0172	0.0326 ± 0.0034	0.432
distal colon	α	0.0256 ± 0.0097	0.0597 ± 0.0164	**0.012**
	β	0.0051 ± 0.0018	0.0428 ± 0.0105	**<0.001**
	γ	0.0067 ± 0.0038	0.0684 ± 0.0227	**0.002**

In the short-term preference test, 3 concentrations of sucrose, MPG+IMP, NaCl, and NaCl+amiloride representing attractive taste stimuli were tested (Figure 5, Tables S4 and S5). The main differences were due to amiloride treatment. However, few additional effects of genotype were recognized (Tables S4 and S5). Accordingly, at least medium and high concentrations of sucrose and MPG+IMP resulted in significantly fewer licks after amiloride intervention (Figure 5A,B), whereas NaCl presentations led to increased licks/water ratios (Figure 5C,D). Licks of water did not show significant changes upon amiloride treatment. Aversive control stimuli (denatonium benzoate and citric acid) were only checked at a single concentration, whereas denatonium benzoate did not result in any differences between the groups and lick responses for citric acid were reduced after amiloride treatment (Figure 5E, Table S4).

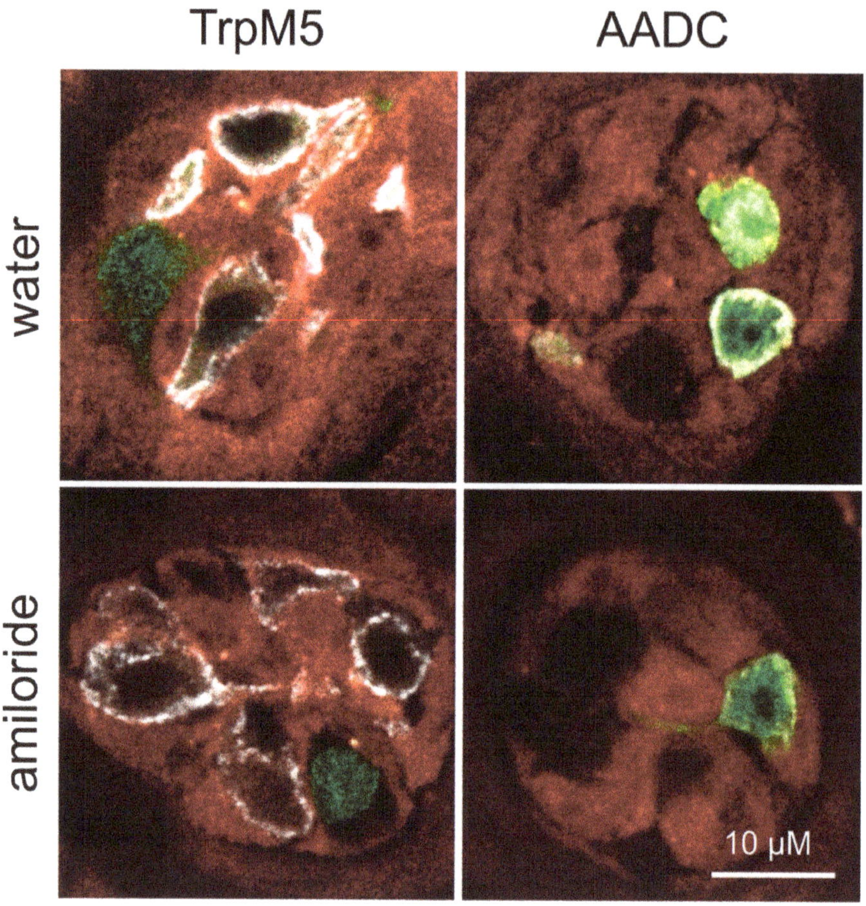

Figure 4. Expression of fluorescent proteins in taste papillae after amiloride intervention. Fungiform papillae sections of Scnn1$^{aa/bb}$ animals expressing GFP (green) and tdRFP (red) fluorescence in αENaC- and βENaC-expressing cells, respectively, were stained for Type II (TrpM5) and Type III (AADC) taste cell markers after amiloride intervention. Therefore, animals received adequate salt diet without or with 300 μM amiloride-containing drinking water prior to sacrifice. Independent of intervention, GFP and tdRFP fluorescence showed no co-localization in taste papillae. Whereas GFP-positive cells always co-expressed AADC, tdRFP-positive cells revealed no overlap with the cell markers TrpM5 or AADC, visualized by immunofluorescence (white). Scale bar applies to all images.

In addition, amiloride treatment did not only change perceived intensity of taste solutions but also increased motivational behavior of the animals. Accordingly, amiloride-treated mice initiated more trials/completed a larger percentage of trials during the 20 min test sessions than did mice that received only water. A trial began with the opening of the shutter and ended 5 s after the mouse made its first lick on the drinking spout (see Section 2). In each session, trials were set to a maximum of 50. Whereas most of the time, animals performed about 20–30 trials per day (e.g., "water-treated" mice, Figure 5 or after dietary intervention, Figure 3), amiloride-treated animals performed about 40–50 trials per session (Figure S2). Despite considerable individual variations for all animals independent of intervention, amiloride-treated animals showed reduced latency, or the time taken to initiate the first lick of a trial, reaching statistical significance for at least one concentration of a presented stimulus. Only for concentrated control stimuli (denatonium benzoate and citric acid) was no or a converse situation recognized, however, it did not reach statistical significance (Figure S2E). Accordingly, the overall performance of amiloride-treated mice was changed.

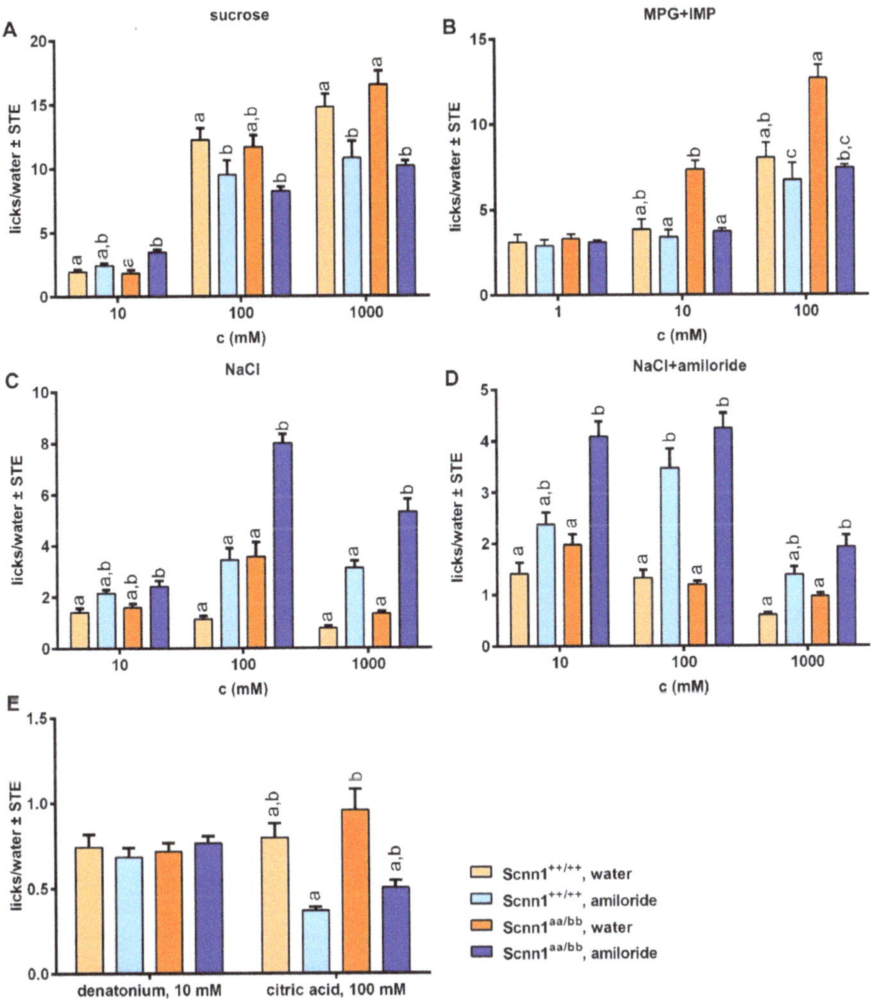

Figure 5. Taste responses of Scnn1$^{++/++}$ and Scnn1$^{aa/bb}$ mice after access to amiloride-containing water. Scnn1$^{++/++}$ and Scnn1$^{aa/bb}$ mice receiving a sodium-adequate diet had either access to 300 μM amiloride-containing water 13 h prior to restriction starting or received water without amiloride. The restriction phase lasted for 22.5 h with access to 2.0 mL water ± 300 μM amiloride and 1 g of food. Lick responses to different concentrated solutions of sucrose (**A**), monopotassium glutamate with inosine 5′monophosphate (MPG+IMP; **B**), sodium chloride (NaCl; **C**), NaCl with amiloride (NaCl+amiloride; **D**), or bitter and sour stimuli (**E**) were determined by an automated gustometer. Each data point represents a mean ± SE of 5 s presentations from 10 to 16 animals tested. Statistical was testing based on UNIANOVA and post-hoc analysis using Bonferroni′s multiple comparison test. Different letters indicate statistical significance.

3.5. Amiloride Interaction with Bitter Taste Receptors

Amiloride was reported to be tasteless to rats and mice at or below 100 μM [41–43], whereas humans perceive bitterness at concentrations above 100 μM [44–47]. If, however, amiloride would also cause bitter perception in mice, this could have an unspecific impact on salt intake in case of synchronous application of amiloride. In order to confirm that amiloride could (not) activate bitter taste receptors, we performed functional expression analyses with mouse and human bitter taste receptor constructs. Transient expression of 25 human and 34 mouse bitter taste receptors revealed that amiloride indeed activated 1 and 7 bitter taste receptors, respectively (Figure 6). The activation of mouse bitter taste receptor Tas2r121 was observed at concentrations of 0.01 mM and above (Figure 6A), whereas the human bitter taste receptors, TAS2R4, TAS2R7, TAS2R13, TAS2R38, TAS2R39, TAS2R43, and TAS2R46 revealed an activation by amiloride at ~100 times higher concentrations (Figure 6B).

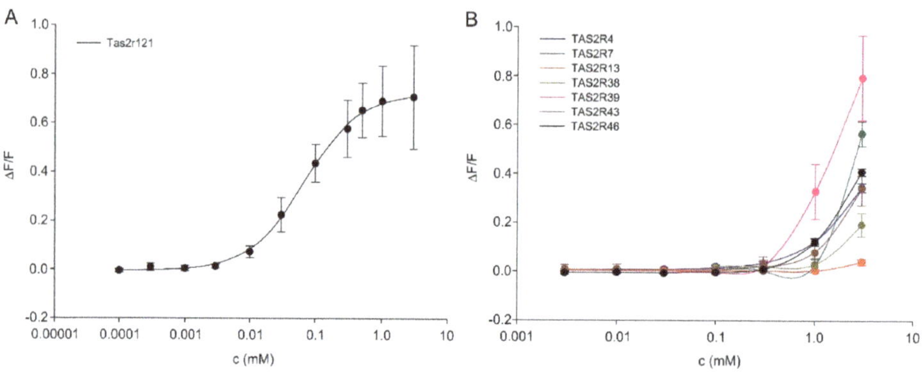

Figure 6. Concentration-response relations of murine (**A**) and human (**B**) bitter taste receptor-expressing cells stimulated with increasing concentrations of amiloride calculated from calcium traces acquired by fluorometric imaging plate reader (FLIPR) recordings. Changes in fluorescence ($\Delta F/F$) were plotted semi-logarithmically versus agonist concentrations.

4. Discussion

Modern society is characterized by high consumption of table salt [48–52], which increases risks of diseases such as stroke, left ventricular hypertrophy, renal stones, or osteoporosis [53–60]. Accordingly, it is of considerable interest to elucidate and understand the molecular and cellular basis of salt taste. In 2010 the relevance of the αENaC subunit in attractive salt taste in mice was confirmed by a conditional knock-out in taste bud cells, impairing amiloride-sensitive salt taste detection, while retaining normal responses to other taste qualities as well as high salt reception [32,61]. This is in agreement with the observation that salt taste is, at least in mice, partially affected by the diuretic drug amiloride, a well-known effector of ENaC in the kidney [9–12]. Based on functional expression experiments in *Xenopus* oocytes, formation of a fully functional ENaC depends on the simultaneous presence of α-, β-, and γ-subunits, even though the α-subunit alone is sufficient to induce weak sodium currents [62–66]. However, a recent study with gene-targeted mice, labeling α- and βENaC-expressing cells by fluorescent proteins, revealed almost no co-localization of the different subunits in taste papillae [33]. To investigate if table salt restriction or overconsumption affects ENaC expression, we used the same α- and βENaC knock-in animals for a dietary intervention study. Over a period of 4 weeks, animals received either an adequate, low, or high salt diet with free access to food and water. Variable sodium content of the diet did not result in any changes in averaged food intake (Figure 1B), supporting the assumption that the caloric need determines the amount of food that is ingested rather than the sodium content. From the time point of dietary change, animals receiving a high salt diet showed a drastic increase in water intake, whereas sodium depletion caused no alterations in water intake in comparison to animals fed an adequate sodium diet (Figure 1C). Accordingly, only ingestion of the high salt diet provided a rapid osmotic stimulation of thirst, as stated before for rats [67,68]. While some reports observed that sodium depletion by low salt diet did not enhance dietary sodium ingestion [69,70], others reported an induction of sodium appetite [71]. Our data suggest the absence of compensatory sodium ingestion. As oral sodium consumption rapidly quenches sodium appetite [72,73], an altered hedonic valance of sodium depending on the body's needs is assumed [71,74,75]. To measure the 'liking' of taste solutions, short-term preference tests were performed (Figure 3). In order to test the wide range of behavioral responses, attractive and aversive restriction conditions were applied. Under conditions favoring attraction behavior, we observed a higher mean lick ratio at 300 mM NaCl in Scnn1$^{aa/bb}$ animals fed with adequate and low salt diet in comparison to the other groups, whereas lick ratios for NaCl in Scnn1$^{aa/bb}$ animals fed with a high salt diet gradually decreased with NaCl concentration (Figure 3C). In comparison, Scnn1$^{++/++}$ mice were relatively indifferent to all NaCl concentrations in comparison to water and did not show concentration-dependent changes in their lick ratio. Under water-restricted (aversive) conditions Scnn1$^{++/++}$ and Scnn1$^{aa/bb}$ animals fed with low salt diet as well as Scnn1$^{aa/bb}$ animals fed with sodium-adequate diet revealed less aversive behavior towards hypertonic saline solutions (300 and 1000 mM) than animals fed with high salt diet or Scnn1$^{++/++}$ animals fed with sodium-adequate diet (Figure 3G). Accordingly, depending on the table salt content of the diet, mice became significantly more adept at avoiding high salt-containing solutions to satisfy the demand for an adequate sodium supply. This is in line with the observation that sodium depletion does neither alter perceived intensity nor quality, assuming that only the hedonic character of table salt is affected [76]. Indeed, sodium-depleted human subjects also displayed an increased preference for high salt diets [77]. Moreover, sodium-depleted rodents drink sodium-containing solutions even at concentrations they would normally reject [18]. Accordingly, sodium deficits seem to trigger an intake behavior towards concentrated sodium solutions.

Furthermore, preference for table salt was reduced by co-application of amiloride without remaining variations between the different diet X genotype constellations under attractive restriction conditions; fitting to the observation that exposure to amiloride diminishes licks to table salt [78,79]. The co-application of amiloride and NaCl under water-restricted conditions had only modest effects on NaCl avoidance (Figure 3H), supporting the view that amiloride-insensitive pathways mediate the detection of high sodium concentrations [80–83].

Conspicuously, next to the diet, expression levels of ENaC subunits also seem to affect NaCl taste. Mice lacking Engrailed-2, a transcription factor critical for neural development, showed an increase in the expression of αENaC subunits in lingual taste papillae, accompanied by increased taste responsivity to 300 mM NaCl and reduced avoidance of salt [84]. Knock-in of fluorescent proteins in the here used Scnn1$^{aa/bb}$ mice also resulted in higher mRNA expression levels of ENaC subunits in taste papillae, but not in non-gustatory lingual tissues (Table S1). These differences in gene expression are probably due to the endowment of both loci with bovine growth hormone polyadenylation signal (BGH) as part of the gene targeting strategy [33]. The presence of the BGH signal results in increased stability and by that mild overexpression of α- and βENaC mRNA [85,86], potentially resulting in the production of higher polypeptide levels from the recombinant locus and eventually accounting for differences between both genotypes (Tables S2 and S3). This is seen, for instance, in the case of the Scnn1$^{aa/bb}$ knock-in mice fed with a sodium-adequate diet, who showed higher lick/water ratios for NaCl (100 to 1000 mM) than Scnn1$^{++/++}$ mice under attractive restriction conditions (Figure 3C). In comparison to these observations, Scnn1$^{aa/bb}$ animals fed with a low salt diet seem more susceptible to the exposure of high NaCl concentrations, resulting in higher lick ratios at 100, 300, and 1000 mM in comparison to corresponding Scnn1$^{++/++}$ animals (Figure 3C). This effect was not seen if NaCl was accompanied by amiloride (Figure 3D). Moreover, it seems that Scnn1$^{aa/bb}$ knock-in mice overexpressing α- and βENaC in general are more sensitive to low MPG+IMP and less sensitive towards high concentrated salt and sour stimuli.

Compared to genotype, either feeding of low or high sodium-containing diets failed to significantly alter relative RNA expression levels of ENaC subunits in gustatory tissue (Table 1). These observations confirm earlier studies, reporting unchanged transcription levels for ENaC subunits in taste buds of sodium-deprived rodents compared with animals fed a control diet [28,36,87,88]. Additionally, no changes in fluorescent protein expression (GFP and tdRFP) were observed, neither regarding cellular localization nor intensity in fungiform papillae. Accordingly, GFP fluorescence was only recognized in Type III cells, whereas tdRFP fluorescence was restricted to non-Type II and -Type III cells (Figure 2). However, with regard to physiologically potentially more relevant tissues for sodium homeostasis, elevated ENaC expression was recognized in the distal colon of animals, which were fed a low salt diet (Table 1). These results confirm and extend earlier studies which reported that low salt diet or application of aldosterone resulted in increased transcription of β- and γENaC subunits in the colon, whereas αENaC subunit transcription remained unchanged [36,89–91]. Some groups further reported that sodium-depleted rats not only showed altered ENaC mRNA levels, but also higher αENaC protein levels in the colon in comparison to animals fed a high salt diet [89,90]. In comparison to that, mice fed a high salt diet showed significantly reduced αENaC expression levels in the kidney in comparison to animals that received a sodium-adequate diet (Table 1). Previously, low salt diet or application of aldosterone was observed to induce the expression of the αENaC subunit in the kidney, accompanied with unchanged expression for β- and γENaC subunits [5,92–96], which was also not observed in this study. Accordingly, compensatory effects for maintaining sodium homeostasis are realized via the colon and kidney rather than taste tissue with regard to adaptations in ENaC expression, indicating the pivotal role of these tissues in sodium reabsorption [97,98].

In addition to dietary intervention, salt appetite in mice was reported to be induced more robustly by exposure to 300 μM amiloride in drinking water over a period of 36 h, as reported recently [38]. Access to 300 μM amiloride-containing drinking water for 36 h resulted in significantly higher expression levels for β- (and nearly also for αENaC) and αENaC in fungiform and posterior lingual papillae, respectively (Table 2). Additionally, once more in the distal colon, all ENaC subunit transcription levels were significantly increased after amiloride treatment (Table 2). Furthermore, lick responses to different taste solutions were altered (Figure 5, Tables S4 and S5). We observed increased lick ratios for NaCl potentially based on a sodium imbalance due to blocked ENaC channels (Figure 5C,D) as well as reduced lick/water ratios for most of the tested concentrations of sucrose and MPG+IMP (Figure 5A,B), indicating perceptual changes and/or reduced attractiveness to the mice. At least for humans, amiloride was recognized to suppress sweet taste [99–101]. Cell culture experiments using a cell line stably expressing human sweet taste receptor revealed that in the presence of 3 mM amiloride responses to sweet tastants like sugars and artificial sweeteners were reduced [99–101]. As sweet and umami taste share molecular and hedonic similarities, comparable effects of long-term amiloride exposure to both taste qualities appear reasonable. Moreover, a reduced lick/water ratio was recognized for 100 mM citric acid (Figure 5E). Whether interactions of amiloride with αENaC, which is expressed in sour-mediating Type III cells, or acid-sensing ion channels (ASICs) might be responsible for this, requires further analyses. At least after short amiloride exposure, neural responses to citric or hydrochloric acid were not affected by amiloride in various species [99,102–104], including the mouse [105]. Otherwise, preliminary results (unpublished) showed that at concentrations above 100 μM amiloride resulted in reduced lick/water ratios in short-term preference tests, indicating that aversive taste perception is triggered. Previously, amiloride has been reported to taste bitter to humans at concentrations above 100 μM [44–47], whereas amiloride was reported to be tasteless to rats and mice at or below 100 μM [41–43]. Our functional expression analysis now identified Tas2r121 as a murine bitter taste receptor for amiloride (Figure 6). This new finding may also have an impact on the observed species differences concerning the amiloride sensitivity of salt taste [44,46,104]. Moreover, the finding that amiloride tastes bitter to mice suggests that amiloride treatment does not only block attraction to low sodium chloride concentrations, but also represents at the same time an aversive taste stimulus. This should be kept in mind when interpreting and planning such experiments.

In summary, this study showed that sodium depletion, feeding a hypertonic saline diet, and amiloride intervention impact taste liking and ENaC expression, with differences regarding subunits and organs. Thereby, colon and kidney seem to be of greater importance to compensate imbalanced sodium homeostasis than gustatory tissue based on the monitored ENaC expression levels. However, effects of genotype were also recognized. As Scnn1$^{aa/bb}$ animals showed higher levels of ENaC subunits (at least at cDNA level) than corresponding wild-type controls, changes of ENaC expression seem to have a prominent impact on taste liking even without dietary interventions. This needs to be addressed in detail in future studies. Additionally, we could confirm that the application of 300 μM amiloride in the drinking water is an efficient way to induce salt appetite. However, amiloride did not only alter taste sensation for salt but also for sucrose, MPG+IMP, and high concentrations of citric acid, indicating a more general influence of this drug and its bitter taste on taste perception.

Supplementary Materials: The following are available online at http://www.mdpi.com/2072-6643/12/4/995/s1, Figure S1: Taste responses of Scnn1$^{++/++}$ and Scnn1$^{aa/bb}$ mice to control stimuli after dietary intervention, Figure S2: Latency to initiate the first lick for different taste stimuli after access to amiloride-containing water, Table S1: Relative expression of ENaC subunits in Scnn1$^{++/++}$ and Scnn1$^{aa/bb}$ mice, Table S2: Statistical significance of different factors on the short-term preference tests of Scnn1$^{++/++}$ and Scnn1$^{aa/bb}$ animals after dietary intervention, Table S3: Statistical significance of different factors on the short-term preference tests of Scnn1$^{++/++}$ and Scnn1$^{aa/bb}$ animals, Table S4: Statistical significance of different factors on the short-term preference tests of Scnn1$^{++/++}$ and Scnn1$^{aa/bb}$ animals after amiloride intervention for 36 h, Table S5: Statistical significance of different factors on the short-term preference tests of Scnn1$^{++/++}$ and Scnn1$^{aa/bb}$ animals after access to amiloride-containing drinking water for 36 h.

Author Contributions: K.L. contributed to study design, performed experiments, data analysis, and interpretation, drafted the manuscript, and prepared the final manuscript version; M.B. provided ideas and contributed to the design of the experiments, data interpretation, drafting the manuscript, and was involved in the preparation of the final manuscript version. W.M. provided conceptual ideas, contributed to the design of the experiments, data interpretation, and preparation of the final manuscript version. All authors have read and agreed to the published version of the manuscript.

Funding: This research was funded by the institutional budget of the German Institute of Human Nutrition provided by the federal ministry for education and research (BMBF) and the ministry for science, research, and culture (MWFK) of the State of Brandenburg, Germany.

Acknowledgments: The authors thank Stefanie Demgensky for excellent technical assistance and Ines Grüner for animal caretaking and experimental support. Additionally, the authors thank Karoline von Websky for providing Power Lab system for the analysis of systolic blood pressure as well as Veit Flockerzi for generously providing the anti-TrpM5 antiserum.

Conflicts of Interest: The authors declare no conflicts of interest.

References

1. Hollenberg, N.K. Surfeit, deficit, and the set point for sodium homeostasis. *Kidney Int.* **1982**, *21*, 883–884. [CrossRef] [PubMed]
2. Damkjaer, M.; Jensen, P.H.; Schwammle, V.; Sprenger, R.R.; Jacobsen, I.A.; Jensen, O.N.; Bie, P. Selective renal vasoconstriction, exaggerated natriuresis and excretion rates of exosomic proteins in essential hypertension. *Acta Physiol. (Oxf.)* **2014**, *212*, 106–118. [CrossRef] [PubMed]
3. Healy, V.; Thompson, C.; Johns, E.J. The adrenergic regulation of proximal tubular Na(+)/H(+) exchanger 3 in the rat. *Acta Physiol. (Oxf.)* **2014**, *210*, 678–689. [CrossRef] [PubMed]
4. Fujita, T. Aldosterone in salt-sensitive hypertension and metabolic syndrome. *J. Mol. Med. (Berl.)* **2008**, *86*, 729–734. [CrossRef]
5. Loffing, J.; Pietri, L.; Aregger, F.; Bloch-Faure, M.; Ziegler, U.; Meneton, P.; Rossier, B.C.; Kaissling, B. Differential subcellular localization of ENaC subunits in mouse kidney in response to high- and low-Na diets. *Am. J. Physiol. Ren. Physiol.* **2000**, *279*, F252–258. [CrossRef]
6. Duc, C.; Farman, N.; Canessa, C.M.; Bonvalet, J.P.; Rossier, B.C. Cell-specific expression of epithelial sodium channel alpha, beta, and gamma subunits in aldosterone-responsive epithelia from the rat: Localization by in situ hybridization and immunocytochemistry. *J. Cell Biol.* **1994**, *127*, 1907–1921. [CrossRef]

7. Kunzelmann, K.; Mall, M. Electrolyte transport in the mammalian colon: Mechanisms and implications for disease. *Physiol. Rev.* **2002**, *82*, 245–289. [CrossRef]
8. Koyama, K.; Sasaki, I.; Naito, H.; Funayama, Y.; Fukushima, K.; Unno, M.; Matsuno, S.; Hayashi, H.; Suzuki, Y. Induction of epithelial Na+ channel in rat ileum after proctocolectomy. *Am. J. Physiol.* **1999**, *276*, G975–984. [CrossRef]
9. Heck, G.L.; Mierson, S.; DeSimone, J.A. Salt taste transduction occurs through an amiloride-sensitive sodium transport pathway. *Science* **1984**, *223*, 403–405. [CrossRef]
10. Brand, J.G.; Teeter, J.H.; Silver, W.L. Inhibition by amiloride of chorda tympani responses evoked by monovalent salts. *Brain Res.* **1985**, *334*, 207–214. [CrossRef]
11. Garty, H.; Palmer, L.G. Epithelial sodium channels: Function, structure, and regulation. *Physiol. Rev.* **1997**, *77*, 359–396. [CrossRef] [PubMed]
12. Kashlan, O.B.; Kleyman, T.R. ENaC structure and function in the wake of a resolved structure of a family member. *Am. J. Physiol. Ren. Physiol.* **2011**, *301*, F684–F696. [CrossRef] [PubMed]
13. Cragoe, E.J., Jr.; Woltersdorf, O.W., Jr.; Bicking, J.B.; Kwong, S.F.; Jones, J.H. Pyrazine diuretics. II. N-amidino-3-amino-5-substituted 6-halopyrazinecarboxamides. *J. Med. Chem.* **1967**, *10*, 66–75. [CrossRef] [PubMed]
14. Cohen, A.B. Hyperkalemic effects of triamterene. *Ann. Intern. Med.* **1966**, *65*, 521–527. [CrossRef] [PubMed]
15. Gombos, E.A.; Freis, E.D.; Moghadam, A. Effects of MK-870 in normal subjects and hypertensive patients. *N. Engl. J. Med.* **1966**, *275*, 1215–1220. [CrossRef] [PubMed]
16. Saha, C.; Eckert, G.J.; Ambrosius, W.T.; Chun, T.Y.; Wagner, M.A.; Zhao, Q.; Pratt, J.H. Improvement in blood pressure with inhibition of the epithelial sodium channel in blacks with hypertension. *Hypertension* **2005**, *46*, 481–487. [CrossRef]
17. Denton, D.A. *The Hunger for Salt*; Springer: Berlin/Heidelberg, Germany; New York, NY, USA, 1882.
18. St John, S.J. The perceptual characteristics of sodium chloride to sodium-depleted rats. *Chem. Senses* **2017**, *42*, 93–103. [CrossRef]
19. Geerling, J.C.; Loewy, A.D. Central regulation of sodium appetite. *Exp. Physiol.* **2008**, *93*, 177–209. [CrossRef]
20. Johnson, A.K.; Thunhorst, R.L. The neuroendocrinology of thirst and salt appetite: Visceral sensory signals and mechanisms of central integration. *Front. Neuroendocrinol.* **1997**, *18*, 292–353. [CrossRef]
21. Fischer, N.M.; Schiefer, C.P.; St John, S.J. Gustatory contributions to sodium appetite: A microstructural analysis. *Chem. Senses* **2016**, *41*, E60.
22. Sato, T.; Miyamoto, T.; Okada, Y. Comparison of gustatory transduction mechanisms in vertebrate taste cells. *Zool. Sci.* **1994**, *11*, 767–780.
23. Lindemann, B. Taste reception. *Physiol. Rev.* **1996**, *76*, 718–766. [CrossRef] [PubMed]
24. Stewart, R.E.; DeSimone, J.A.; Hill, D.L. New perspectives in a gustatory physiology: Transduction, development, and plasticity. *Am. J. Physiol.* **1997**, *272*, C1–26. [CrossRef] [PubMed]
25. Benos, D.J. Amiloride: A molecular probe of sodium transport in tissues and cells. *Am. J. Physiol.* **1982**, *242*, C131–145. [CrossRef]
26. Lindemann, B. Receptors and transduction in taste. *Nature* **2001**, *413*, 219–225. [CrossRef]
27. Ninomiya, Y.; Funakoshi, M. Amiloride inhibition of responses of rat single chorda tympani fibers to chemical and electrical tongue stimulations. *Brain Res.* **1988**, *451*, 319–325. [CrossRef]
28. Lin, W.; Finger, T.E.; Rossier, B.C.; Kinnamon, S.C. Epithelial Na+ channel subunits in rat taste cells: Localization and regulation by aldosterone. *J. Comp. Neurol.* **1999**, *405*, 406–420. [CrossRef]
29. Gilbertson, T.A.; Kinnamon, S.C. Making sense of chemicals. *Chem. Biol.* **1996**, *3*, 233–237. [CrossRef]
30. Boughter, J.D., Jr.; Gilbertson, T.A. From channels to behavior: An integrative model of NaCl taste. *Neuron* **1999**, *22*, 213–215. [CrossRef]
31. Kretz, O.; Barbry, P.; Bock, R.; Lindemann, B. Differential expression of RNA and protein of the three pore-forming subunits of the amiloride-sensitive epithelial sodium channel in taste buds of the rat. *J. Histochem. Cytochem. Off. J. Histochem. Soc.* **1999**, *47*, 51–64. [CrossRef]
32. Chandrashekar, J.; Kuhn, C.; Oka, Y.; Yarmolinsky, D.A.; Hummler, E.; Ryba, N.J.; Zuker, C.S. The cells and peripheral representation of sodium taste in mice. *Nature* **2010**, *464*, 297–301. [CrossRef] [PubMed]
33. Lossow, K.; Hermans-Borgmeyer, I.; Meyerhof, W.; Behrens, M. Segregated expression of ENaC subunits in taste cells. *Chem. Senses* **2020**. [CrossRef] [PubMed]

34. Jasti, J.; Furukawa, H.; Gonzales, E.B.; Gouaux, E. Structure of acid-sensing ion channel 1 at 1.9 A resolution and low pH. *Nature* **2007**, *449*, 316–323. [CrossRef] [PubMed]
35. Noreng, S.; Bharadwaj, A.; Posert, R.; Yoshioka, C.; Baconguis, I. Structure of the human epithelial sodium channel by cryo-electron microscopy. *Elife* **2018**, 7. [CrossRef]
36. Huang, T.; Stahler, F. Effects of dietary Na+ deprivation on epithelial Na+ channel (ENaC), BDNF, and TrkB mRNA expression in the rat tongue. *Bmc Neurosci.* **2009**, *10*, 19. [CrossRef]
37. Hedrich, H. *The Laboratory Mouse*, 1st ed.; Elsevier, Academic Press: Cambridge, MA, USA, 2004.
38. Raymond, M.A.; Chowdhury, T.; Mast, T.G.; Breza, J.M. A simple method for a rapid induction of salt appetite in mice. *Annu. Meet. Assoc. Chemorecept. Sci.* **2017**. [CrossRef]
39. Lossow, K.; Hubner, S.; Roudnitzky, N.; Slack, J.P.; Pollastro, F.; Behrens, M.; Meyerhof, W. Comprehensive analysis of mouse bitter taste receptors reveals different molecular receptive ranges for orthologous receptors in mice and humans. *J. Biol. Chem.* **2016**, *291*, 15358–15377. [CrossRef]
40. Kusumakshi, S.; Voigt, A.; Hubner, S.; Hermans-Borgmeyer, I.; Ortalli, A.; Pyrski, M.; Dorr, J.; Zufall, F.; Flockerzi, V.; Meyerhof, W.; et al. A binary genetic approach to characterize trpm5 cells in mice. *Chem. Senses* **2015**, *40*, 413–425. [CrossRef]
41. Hill, D.L.; Formaker, B.K.; White, K.S. Perceptual characteristics of the amiloride-suppressed sodium chloride taste response in the rat. *Behav. Neurosci.* **1990**, *104*, 734–741. [CrossRef]
42. Markison, S.; Spector, A.C. Amiloride is an ineffective conditioned stimulus in taste aversion learning. *Chem. Senses* **1995**, *20*, 559–563. [CrossRef]
43. Eylam, S.; Tracy, T.; Garcea, M.; Spector, A.C. Amiloride is an ineffective conditioned stimulus in taste aversion learning in C57BL/6J and DBA/2J mice. *Chem. Senses* **2003**, *28*, 681–689. [CrossRef] [PubMed]
44. Desor, J.A.; Finn, J. Effects of amiloride on salt taste in humans. *Chem. Senses* **1989**, *14*, 793–803. [CrossRef]
45. McCutcheon, N.B. Human psychophysical studies of saltiness suppression by amiloride. *Physiol. Behav.* **1992**, *51*, 1069–1074. [CrossRef]
46. Smith, D.V.; Ossebaard, C.A. Amiloride suppression of the taste intensity of sodium chloride: Evidence from direct magnitude scaling. *Physiol. Behav.* **1995**, *57*, 773–777. [CrossRef]
47. Halpern, B.P. Amiloride and vertebrate gustatory responses to NaCl. *Neurosci. Biobehav. Rev.* **1998**, *23*, 5–47. [CrossRef]
48. Brown, I.J.; Tzoulaki, I.; Candeias, V.; Elliott, P. Salt intakes around the world: Implications for public health. *Int. J. Epidemiol.* **2009**, *38*, 791–813. [CrossRef]
49. Mozaffarian, D.; Fahimi, S.; Singh, G.M.; Micha, R.; Khatibzadeh, S.; Engell, R.E.; Lim, S.; Danaei, G.; Ezzati, M.; Powles, J.; et al. Global sodium consumption and death from cardiovascular causes. *N. Engl. J. Med.* **2014**, *371*, 624–634. [CrossRef]
50. O'Donnell, M.; Mente, A.; Rangarajan, S.; McQueen, M.J.; Wang, X.; Liu, L.; Yan, H.; Lee, S.F.; Mony, P.; Devanath, A.; et al. Urinary sodium and potassium excretion, mortality, and cardiovascular events. *N. Engl. J. Med.* **2014**, *371*, 612–623. [CrossRef]
51. WHO. *Guideline: Sodium Intake for Adults and Children*; World Health Organization (WHO): Geneva, Switzerland, 2012; Available online: www.who.int (accessed on 1 October 2018).
52. Meneton, P.; Jeunemaitre, X.; de Wardener, H.E.; MacGregor, G.A. Links between dietary salt intake, renal salt handling, blood pressure, and cardiovascular diseases. *Physiol. Rev.* **2005**, *85*, 679–715. [CrossRef]
53. He, F.J.; MacGregor, G.A. Reducing population salt intake worldwide: From evidence to implementation. *Prog. Cardiovasc. Dis.* **2010**, *52*, 363–382. [CrossRef]
54. He, F.J.; MacGregor, G.A. A comprehensive review on salt and health and current experience of worldwide salt reduction programmes. *J. Hum. Hypertens.* **2009**, *23*, 363–384. [CrossRef] [PubMed]
55. Bibbins-Domingo, K.; Chertow, G.M.; Coxson, P.G.; Moran, A.; Lightwood, J.M.; Pletcher, M.J.; Goldman, L. Projected effect of dietary salt reductions on future cardiovascular disease. *N. Engl. J. Med.* **2010**, *362*, 590–599. [CrossRef] [PubMed]
56. Sacks, F.M.; Campos, H. Dietary therapy in hypertension. *N. Engl. J. Med.* **2010**, *362*, 2102–2112. [CrossRef] [PubMed]
57. Tsugane, S.; Sasazuki, S.; Kobayashi, M.; Sasaki, S. Salt and salted food intake and subsequent risk of gastric cancer among middle-aged Japanese men and women. *Br. J. Cancer* **2004**, *90*, 128–134. [CrossRef] [PubMed]

58. Devine, A.; Criddle, R.A.; Dick, I.M.; Kerr, D.A.; Prince, R.L. A longitudinal study of the effect of sodium and calcium intakes on regional bone density in postmenopausal women. *Am. J. Clin. Nutr.* **1995**, *62*, 740–745. [CrossRef] [PubMed]
59. Asaria, P.; Chisholm, D.; Mathers, C.; Ezzati, M.; Beaglehole, R. Chronic disease prevention: Health effects and financial costs of strategies to reduce salt intake and control tobacco use. *Lancet* **2007**, *370*, 2044–2053. [CrossRef]
60. Strazzullo, P.; D'Elia, L.; Kandala, N.B.; Cappuccio, F.P. Salt intake, stroke, and cardiovascular disease: Meta-analysis of prospective studies. *BMJ* **2009**, *339*, b4567. [CrossRef]
61. Bosak, N.P.; Inoue, M.; Nelson, T.M.; Hummler, E.; Ishiwatari, Y.; Bachmanov, A.A. Epithelial sodium channel (ENaC) is involved in reception of sodium taste: Evidence from mice with a tissue-specific conditional targeted mutation of the ENaC gene (abstract), thirty-second annual meeting of the association for chemoreception sciences. *Chem. Senses* **2010**, *35*, A2.
62. Canessa, C.M.; Schild, L.; Buell, G.; Thorens, B.; Gautschi, I.; Horisberger, J.D.; Rossier, B.C. Amiloride-sensitive epithelial Na+ channel is made of three homologous subunits. *Nature* **1994**, *367*, 463–467. [CrossRef]
63. McDonald, F.J.; Price, M.P.; Snyder, P.M.; Welsh, M.J. Cloning and expression of the beta- and gamma-subunits of the human epithelial sodium channel. *Am. J. Physiol.* **1995**, *268*, C1157–C1163. [CrossRef]
64. Giraldez, T.; Afonso-Oramas, D.; Cruz-Muros, I.; Garcia-Marin, V.; Pagel, P.; Gonzalez-Hernandez, T.; Alvarez de la Rosa, D. Cloning and functional expression of a new epithelial sodium channel delta subunit isoform differentially expressed in neurons of the human and monkey telencephalon. *J. Neurochem.* **2007**, *102*, 1304–1315. [CrossRef] [PubMed]
65. Edelheit, O.; Hanukoglu, I.; Dascal, N.; Hanukoglu, A. Identification of the roles of conserved charged residues in the extracellular domain of an epithelial sodium channel (ENaC) subunit by alanine mutagenesis. *Am. J. Physiol. Ren. Physiol.* **2011**, *300*, F887–897. [CrossRef] [PubMed]
66. Edelheit, O.; Ben-Shahar, R.; Dascal, N.; Hanukoglu, A.; Hanukoglu, I. Conserved charged residues at the surface and interface of epithelial sodium channel subunits–roles in cell surface expression and the sodium self-inhibition response. *Febs J.* **2014**, *281*, 2097–2111. [CrossRef] [PubMed]
67. Stricker, E.M.; Hoffmann, M.L.; Riccardi, C.J.; Smith, J.C. Increased water intake by rats maintained on high NaCl diet: Analysis of ingestive behavior. *Physiol. Behav.* **2003**, *79*, 621–631. [CrossRef]
68. Ramsey, D.J.; Thrasher, T.N. Thirst and water balance. Neurobiology of food and fluid intake. In *Handbook of Behavioral Neurobiology*; Stricker, E.M., Ed.; Plenum Press: New York, NY, USA, 1990; Volume 10, pp. 353–386.
69. Contreras, R.J.; Hatton, G.I. Gustatory adaptation as an explanation for dietary-induced sodium appetite. *Physiol. Behav.* **1975**, *15*, 569–576. [CrossRef]
70. Stricker, E.M.; Thiels, E.; Verbalis, J.G. Sodium appetite in rats after prolonged dietary sodium deprivation: A sexually dimorphic phenomenon. *Am. J. Physiol.* **1991**, *260*, R1082–R1088. [CrossRef]
71. Berridge, K.C.; Flynn, F.W.; Schulkin, J.; Grill, H.J. Sodium depletion enhances salt palatability in rats. *Behav. Neurosci.* **1984**, *98*, 652–660. [CrossRef]
72. Nachman, M.; Valentino, D.A. Roles of taste and postingestional factors in the satiation of sodium appetite in rats. *J. Comp. Physiol. Psychol.* **1966**, *62*, 280–283. [CrossRef]
73. Wolf, G.; Schulkin, J.; Simson, P.E. Multiple factors in the satiation of salt appetite. *Behav. Neurosci.* **1984**, *98*, 661–673. [CrossRef]
74. Handal, P.J. Immediate acceptance of sodium salts by sodium deficient rats. *Psychon. Sci.* **1965**, *3*, 315–316. [CrossRef]
75. Fortin, S.M.; Roitman, M.F. Physiological state tunes mesolimbic signaling: Lessons from sodium appetite and inspiration from Randall R. Sakai. *Physiol. Behav.* **2017**, *178*, 21–27. [CrossRef] [PubMed]
76. St John, S.J.; McBrayer, A.M.; Krauskopf, E.E. Sodium carbonate is saltier than sodium chloride to sodium-depleted rats. *Chem. Senses* **2017**, *42*, 647–653. [CrossRef] [PubMed]
77. Beauchamp, G.K.; Bertino, M.; Burke, D.; Engelman, K. Experimental sodium depletion and salt taste in normal human volunteers. *Am. J. Clin. Nutr.* **1990**, *51*, 881–889. [CrossRef] [PubMed]
78. Bernstein, I.L.; Hennessy, C.J. Amiloride-sensitive sodium channels and expression of sodium appetite in rats. *Am. J. Physiol.* **1987**, *253*, R371–374. [CrossRef]
79. McCutcheon, N.B. Sodium deficient rats are unmotivated by sodium chloride solutions mixed with the sodium channel blocker amiloride. *Behav. Neurosci.* **1991**, *105*, 764–766. [CrossRef]

80. Oka, Y.; Butnaru, M.; von Buchholtz, L.; Ryba, N.J.; Zuker, C.S. High salt recruits aversive taste pathways. *Nature* **2013**, *494*, 472–475. [CrossRef]
81. Ren, Z.; Rhyu, M.R.; Phan, T.H.; Mummalaneni, S.; Murthy, K.S.; Grider, J.R.; DeSimone, J.A.; Lyall, V. TRPM5-dependent amiloride- and benzamil-insensitive NaCl chorda tympani taste nerve response. *Am. J. Physiol. Gastrointest. Liver Physiol.* **2013**, *305*, G106–117. [CrossRef]
82. Lewandowski, B.C.; Sukumaran, S.K.; Margolskee, R.F.; Bachmanov, A.A. Amiloride-insensitive salt taste is mediated by two populations of type iii taste cells with distinct transduction mechanisms. *J. Neurosci. Off. J. Soc. Neurosci.* **2016**, *36*, 1942–1953. [CrossRef]
83. Zhang, Y.; Hoon, M.A.; Chandrashekar, J.; Mueller, K.L.; Cook, B.; Wu, D.; Zuker, C.S.; Ryba, N.J. Coding of sweet, bitter, and umami tastes: Different receptor cells sharing similar signaling pathways. *Cell* **2003**, *112*, 293–301. [CrossRef]
84. Gupta, A.; Li, X.; DiCicco-Bloom, E.; Bello, N.T. Altered salt taste response and increased tongue epithelium Scnna1 expression in adult Engrailed-2 null mice. *Physiol. Behav.* **2018**, *194*, 410–419. [CrossRef]
85. Ribeiro, S.C.; Monteiro, G.A.; Prazeres, D.M. The role of polyadenylation signal secondary structures on the resistance of plasmid vectors to nucleases. *J. Gene Med.* **2004**, *6*, 565–573. [CrossRef] [PubMed]
86. Azzoni, A.R.; Ribeiro, S.C.; Monteiro, G.A.; Prazeres, D.M. The impact of polyadenylation signals on plasmid nuclease-resistance and transgene expression. *J. Gene Med.* **2007**, *9*, 392–402. [CrossRef] [PubMed]
87. Lindemann, B.; Barbry, P.; Kretz, O.; Bock, R. Occurrence of ENaC subunit mRNA and immunocytochemistry of the channel subunits in taste buds of the rat vallate papilla. *Ann. N. Y. Acad. Sci.* **1998**, *855*, 116–127. [CrossRef] [PubMed]
88. Shigemura, N.; Islam, A.A.; Sadamitsu, C.; Yoshida, R.; Yasumatsu, K.; Ninomiya, Y. Expression of amiloride-sensitive epithelial sodium channels in mouse taste cells after chorda tympani nerve crush. *Chem. Senses* **2005**, *30*, 531–538. [CrossRef]
89. Lingueglia, E.; Renard, S.; Waldmann, R.; Voilley, N.; Champigny, G.; Plass, H.; Lazdunski, M.; Barbry, P. Different homologous subunits of the amiloride-sensitive Na+ channel are differently regulated by aldosterone. *J. Biol. Chem.* **1994**, *269*, 13736–13739.
90. Renard, S.; Voilley, N.; Bassilana, F.; Lazdunski, M.; Barbry, P. Localization and regulation by steroids of the alpha, beta and gamma subunits of the amiloride-sensitive Na+ channel in colon, lung and kidney. *Pflug. Arch. Eur. J. Physiol.* **1995**, *430*, 299–307. [CrossRef] [PubMed]
91. Asher, C.; Wald, H.; Rossier, B.C.; Garty, H. Aldosterone-induced increase in the abundance of Na+ channel subunits. *Am. J. Physiol.* **1996**, *271*, C605–C611. [CrossRef]
92. Ono, S.; Kusano, E.; Muto, S.; Ando, Y.; Asano, Y. A low-Na+ diet enhances expression of mRNA for epithelial Na+ channel in rat renal inner medulla. *Pflug. Arch. Eur. J. Physiol.* **1997**, *434*, 756–763. [CrossRef]
93. Stokes, J.B.; Sigmund, R.D. Regulation of rENaC mRNA by dietary NaCl and steroids: Organ, tissue, and steroid heterogeneity. *Am. J. Physiol.* **1998**, *274*, C1699–C1707. [CrossRef]
94. Masilamani, S.; Kim, G.H.; Mitchell, C.; Wade, J.B.; Knepper, M.A. Aldosterone-mediated regulation of ENaC alpha, beta, and gamma subunit proteins in rat kidney. *J. Clin. Investig.* **1999**, *104*, R19–R23. [CrossRef]
95. MacDonald, P.; MacKenzie, S.; Ramage, L.E.; Seckl, J.R.; Brown, R.W. Corticosteroid regulation of amiloride-sensitive sodium-channel subunit mRNA expression in mouse kidney. *J. Endocrinol.* **2000**, *165*, 25–37. [CrossRef] [PubMed]
96. Frindt, G.; Palmer, L.G. Regulation of epithelial Na+ channels by adrenal steroids: Mineralocorticoid and glucocorticoid effects. *Am. J. Physiol. Ren. Physiol.* **2012**, *302*, F20–F26. [CrossRef]
97. Malsure, S.; Wang, Q.; Charles, R.P.; Sergi, C.; Perrier, R.; Christensen, B.M.; Maillard, M.; Rossier, B.C.; Hummler, E. Colon-specific deletion of epithelial sodium channel causes sodium loss and aldosterone resistance. *J. Am. Soc. Nephrol. JASN* **2014**, *25*, 1453–1464. [CrossRef] [PubMed]
98. Marunaka, Y.; Marunaka, R.; Sun, H.; Yamamoto, T.; Kanamura, N.; Taruno, A. Na(+) homeostasis by epithelial Na(+) channel (ENaC) and Nax channel (Nax): Cooperation of ENaC and Nax. *Ann. Transl. Med.* **2016**, *4*, S11. [CrossRef]
99. Schiffman, S.S.; Lockhead, E.; Maes, F.W. Amiloride reduces the taste intensity of Na+ and Li+ salts and sweeteners. *Proc. Natl. Acad. Sci. USA* **1983**, *80*, 6136–6140. [CrossRef] [PubMed]
100. Imada, T.; Misaka, T.; Fujiwara, S.; Okada, S.; Fukuda, Y.; Abe, K. Amiloride reduces the sweet taste intensity by inhibiting the human sweet taste receptor. *Biochem. Biophys. Res. Commun.* **2010**, *397*, 220–225. [CrossRef]

101. Yee, K.K.; Sukumaran, S.K.; Kotha, R.; Gilbertson, T.A.; Margolskee, R.F. Glucose transporters and ATP-gated K+ (KATP) metabolic sensors are present in type 1 taste receptor 3 (T1r3)-expressing taste cells. *Proc. Natl. Acad. Sci. USA* **2011**, *108*, 5431–5436. [CrossRef]
102. Hellekant, G.; DuBois, G.E.; Roberts, T.W.; van der Wel, H. On the gustatory effect of amiloride in the monkey (Macaca mulatto) *Chem. Senses* **1988**, *13*, 89–93. [CrossRef]
103. DeSimone, J.A.; Callaham, E.M.; Heck, G.L. Chorda tympani taste response of rat to hydrochloric acid subject to voltage-clamped lingual receptive field. *Am. J. Physiol.* **1995**, *268*, C1295–C1300. [CrossRef]
104. Ossebaard, C.A.; Smith, D.V. Effect of amiloride on the taste of NaCl, Na-gluconate and KCl in humans: Implications for Na+ receptor mechanisms. *Chem. Senses* **1995**, *20*, 37–46. [CrossRef]
105. Miyamoto, T.; Fujiyama, R.; Okada, Y.; Sato, T. Sour transduction involves activation of NPPB-sensitive conductance in mouse taste cells. *J. Neurophysiol.* **1998**, *80*, 1852–1859. [CrossRef] [PubMed]

nutrients

MDPI

Article

Kokumi Taste Active Peptides Modulate Salt and Umami Taste

Mee-Ra Rhyu [1,*], Ah-Young Song [1], Eun-Young Kim [1], Hee-Jin Son [1], Yiseul Kim [1], Shobha Mummalaneni [2], Jie Qian [2], John R. Grider [2] and Vijay Lyall [2,*]

1 Korea Food Research Institute, Wanju-gun, Jeollabuk-do 55365, Korea; ahsong47@naver.com (A.-Y.S.); keykos@naver.com (E.-Y.K.); thsgmlwls77@naver.com (H.-J.S.); kimys@kfri.re.kr (Y.K.)

2 Department of Physiology and Biophysics, Virginia Commonwealth University, Richmond, VA 23298, USA; shobha.mummalaneni@vcuhealth.org (S.M.); jie.qian@vcuhealth.org (J.Q.); john.grider@vcuhealth.org (J.R.G.)

* Correspondence: mrrhyu@kfri.re.kr (M.-R.R.); vijay.lyall@vcuhealth.org (V.L.); Tel.: +82-63-219-9268 (M.-R.R.); +1-(804)-828-9759 (V.L.); Fax: +82-63-219-9876 (M.-R.R.); +1-(804)-827-0947 (V.L.)

Received: 31 March 2020; Accepted: 22 April 2020; Published: 24 April 2020

check for updates

Abstract: Kokumi taste substances exemplified by γ-glutamyl peptides and Maillard Peptides modulate salt and umami tastes. However, the underlying mechanism for their action has not been delineated. Here, we investigated the effects of a kokumi taste active and inactive peptide fraction (500–10,000 Da) isolated from mature (FII_m) and immature (FII_{im}) Ganjang, a typical Korean soy sauce, on salt and umami taste responses in humans and rodents. Only FII_m (0.1–1.0%) produced a biphasic effect in rat chorda tympani (CT) taste nerve responses to lingual stimulation with 100 mM NaCl + 5 μM benzamil, a specific epithelial Na^+ channel blocker. Both elevated temperature (42 °C) and FII_m produced synergistic effects on the NaCl + benzamil CT response. At 0.5% FII_m produced the maximum increase in rat CT response to NaCl + benzamil, and enhanced salt taste intensity in human subjects. At 2.5% FII_m enhanced rat CT response to glutamate that was equivalent to the enhancement observed with 1 mM IMP. In human subjects, 0.3% FII_m produced enhancement of umami taste. These results suggest that FII_m modulates amiloride-insensitive salt taste and umami taste at different concentration ranges in rats and humans.

Keywords: Korean soy sauce; kokumi; umami; salty; chorda tympani; amiloride-insensitive salt taste pathway

1. Introduction

Mammals use G-protein-coupled receptors (GPCRs) expressed in Type II taste receptor cells (TRCs) to detect bitter, sweet, and umami taste stimuli. While amiloride-sensitive salt taste is detected by Type 1 cells expressing epithelial Na^+ channels, Type II and Type III cells mediate amiloride-insensitive salt taste. Otopetrin-1 proton selective channel expressed in Type III TRCs detects sour taste stimuli [1–6]. Much progress has been made in the identification of taste receptors and the downstream signalling mechanisms involved in the transduction of salty, sour, sweet, bitter and umami taste qualities. However, psychophysical, neural, and cellular studies have long suggested that cell to cell interactions within taste buds and interactions between different taste receptors enhance or suppress taste responses [7,8]. The synergism between monosodium glutamate (MSG) and 5′-ribonucleotides, a distinct characteristic of umami taste, is an example of a binary taste interaction between agonists [9,10]. Additionally, umami peptides modulate bitterness by interfering with ligand binding to the human bitter taste receptor TAS2R16 [11]. Interactions between non-tastants and tastants can also modulate taste intensity.

SE-1, a sweet receptor positive allosteric modulator, binds to the sweet receptor without activating it, but does so in a manner that causes the orthogonal ligands to bind with higher affinity [12,13].

Kokumi taste has the characteristics of enhancing continuity, thickness, and mouthfeel, and was first observed in an aqueous extract of garlic in an umami solution. Kokumi produces its effect despite minimally eliciting any taste on its own [14]. Sulfur-containing compounds and their γ-glutamyl peptides, including γ-Glu-Cys-Gly (GSH) were suggested to be kokumi-active substances [14–17]. Because GSH was identified as an endogenous modulator of the calcium-sensing receptor, which participates in calcium homeostasis in the body [18], identification of GSH as an active component suggests the involvement of calcium-sensing receptor in kokumi perception [19]. Subsequent sensory analyses using various extracellular calcium-sensing receptor agonists have shown that kokumi did have a taste-enhancing effect on sweet, salty, and umami taste [19]. Not only does the γ-glutamyl peptide elicit kokumi taste, but the Maillard [20] reacted peptides (MRPs), which are gradually formed by longer maturation of Korean soy sauce, Ganjang (JGN), have been suggested to play a role in the kokumi taste in humans [21]. JGN is generally stored at ambient conditions for a year, and for up to four years or more to attain full maturity. The taste characteristics of kokumi increase as the maturation progresses.

Salt taste is detected by at least two receptor-mediated pathways. One pathway is Na^+ specific and involves Na^+ influx into TRCs that express amiloride- and benzamil (Bz)-sensitive epithelial Na^+ channels (ENaCs) [22,23]. The second pathway is amiloride-insensitive and is cation nonselective, and does not discriminate between Na^+, K^+ and $NH_4{}^+$ salts. The contribution of these two pathways varies in different taste receptive fields. Approximately 65% of TRCs in the fungiform taste buds exhibit functional ENaCs, 35% of TRCs in foliate taste buds are amiloride-sensitive, while TRCs in the circumvallate are completely amiloride-insensitive, and do not seem to express functional ENaCs [24]. Although at present the identity of the amiloride-insensitive Na^+ pathways in TRCs remains elusive, the amiloride- and Bz-insensitive salt taste receptors are the predominant transducers of salt taste in humans [25–27].

Investigations conducted using the Maillard reaction between peptides (1000–5000 Da) isolated from soy protein hydrolysate and xylose (Xyl-MRPs) have been known to enhance umami, continuity, and mouthfeel in umami solution, support the notion that MRPs are another class of kokumi substances [28]. Interestingly, Xyl-MRPs not only modulate umami taste, but also modulate salt taste. The effect of Xyl-MRPs on salt taste is observed at much lower concentrations than those that increase the umami taste [27]. Over a range of concentrations, Xyl-MRPs [27,29] reversibly enhanced the Bz-insensitive NaCl chorda tympani (CT) taste nerve response in rodents, whereas, at high concentrations, they inhibited the Bz-insensitive NaCl CT response. The effect of Xyl-MRPs on the Bz-insensitive NaCl CT responses were transient receptor potential vanilloid 1 (TRPV1)-dependent. In human sensory evaluation, at low salt concentrations, galacturonic acid MRPs (GalA-MRPs) [27] enhanced human salt taste perception. These data suggest that, in both rodents and humans, MRPs induce changes in amiloride-insensitive salt taste and umami taste.

In this paper, we investigated the effects of a naturally occurring MRPs fraction (500–10,000 Da, FII) isolated from mature (FII_m; 4-year old) and immature (FII_{im}; 1-year old) JGN on salty and umami taste responses in rodents and human subjects. Effects of FII_m and FII_{im} were investigated on the Bz-insensitive NaCl CT responses and their interactions with TRPV1 modulators, and glutamate CT responses in rats. Effects of FII_m and FII_{im} were investigated on behavioral responses to NaCl in C57BL/6 mice, and on the sensory evaluation of salty and umami tastes in human subjects. Our results demonstrate that FII_m produces concentration-dependent biphasic effects on amiloride-insensitive neural and behavioral responses to NaCl in rodents. Above the concentrations that modulate salty taste, FII_m enhanced CT responses to glutamate. In human subjects, FII_m produced concentration-dependent biphasic effects on salt taste perception and at higher concentrations enhanced umami taste. These results suggest that FII_m modulates salty and umami taste in rodents and humans via similar mechanisms.

2. Materials and Methods

2.1. Isolation of FII_m and FII_{im} from JGN

FII fraction containing MRPs of molecular weight (MW) ranging between 500 and 10,000 Da was isolated from immature (FII_{im}; 1-year old) and mature (FII_m; 4-year old) JGN with an ultra-filtration unit (Model 840, Amicon Inc., Beverly MA, USA) using YM-10 (MW cutoff 10,000 Da) and YC-05 (MW cutoff 500 Da) membranes (Millipore Co., Bedford, MA, USA) at 2–4 °C under N_2 pressure. Each fraction was lyophilized and stored in a desiccated freezer at −20 °C until use. FII_m was further separated using YM5, YM3 or YM1 Millipore membranes that had a cut off MW of 5000, 3000 and 1000 Da, respectively. These fractions were: FII_{ma} (MW 500–1000 Da), FII_{mb} (MW 1000–3000 Da), FII_{mc} (MW 3000–5000 Da) and FII_{md} (MW 5000–10,000 Da). FII_m and FII_{im} are the unfractionated MRPs and $FII_{m(a\text{-}d)}$ are the sub-fractions of different molecular weight. Successive column chromatography was performed with FII_m to obtain aromatic, basic, acidic, and neutral conjugated peptide fractions using activated charcoal (60 cm long and 4.0 cm I.D.; Junsei Chemical Co. Ltd., Tokyo, Japan), cation-exchanger (60 cm long and 3.0 cm I.D.; Amberlite IRC-50), and anion-exchanger (60 cm long and 3.0 cm I.D.; Amberlite IRA 400, both from Sigma Co. Ltd., St. Louis, MO, USA) [30,31].

2.2. CT Taste Nerve Recordings

In contrast to glossopharyngeal nerve response to NaCl, the predominant NaCl CT response in rodents is amiloride sensitive. However, a significant part of the NaCl CT response is Bz- and amiloride insensitive across the concentration-response range of NaCl [32]. The identity of the amiloride-insensitive receptor presently at best remains elusive in the circumvallate taste receptive field. Our previous studies suggest that in the anterior tongue the amiloride-insensitive pathway is a non-selective cation channel that is sensitive to resiniferatoxin (RTX), N-(3-methoxyphenyl)-4-chlorocinnamide, SB-366791 (SB), capsazepine, iodo-RTX, and temperature [33,34]. We have previously investigated the effect of various salt taste modulators on the Bz-insensitive NaCl CT response using both rats and mice [27,29,32–36]. To compare the results of the effects of FII_m on neural responses to NaCl with previously published results, we monitored Bz-insensitive NaCl CT response in rats.

Animals were housed in the Virginia Commonwealth University (VCU) animal facility in accordance with institutional guidelines. All animal protocols were approved by the Institutional Animal Care and Use Committee (IACUC #AD20116). Female Sprague-Dawley rats (150–200 gm) were anesthetized by intraperitoneal injection of pentobarbital (60 mg/Kg) and supplemental pentobarbital (20 mg/Kg) was administered as necessary to maintain surgical anesthesia. The animal's corneal reflex and toe-pinch reflex were used to monitor the depth of surgical anesthesia. Body temperatures were maintained at 37 °C with a Deltaphase Isothermal PAD (Model 39 DP: Braintree Scientific Inc. Braintree, MA, USA). The left CT nerve was exposed laterally as it exited the tympanic bulla and placed onto a 32G platinum/iridium wire electrode. CT responses were recorded and analyzed as described previously [27,29,32–36].

The composition of rinse and NaCl stimulating solutions is shown in Table 1. CT responses in rats were monitored while the anterior lingual surface was stimulated first with the rinse solution (R) and then with salt solutions containing 0.1–0.5% FII_{im}, FII_m and FII_m sub-fractions ($FII_{m(a\text{–}d)}$). The pH of the rinse solution and the salt solutions was adjusted to 6.1. In some experiments Bz (5 μM) was added to salt solutions to block Na^+ entry into TRCs through apical epithelial Bz-sensitive ENaCs. CT responses were also recorded at 23 °C and 42 °C. In additional experiments we tested the effect of FII_m on the CT response to MSG and MSG + 5′-inosine monophosphate (IMP), a specific modulator of umami taste [37]. CT responses to MSG were monitored in the presence of Bz to eliminate the contribution of Na^+ to the glutamate CT response [33] and SB (1 μM), a TRPV1 blocker [38]. In CT experiments the tonic (steady-state) part of the NaCl CT response or glutamate CT response was quantified and normalized to CT responses to 0.3M NH_4Cl. The normalized data were reported as the mean (M) ± SEM of the number of animals (n). Student's t-test was employed to analyze the differences between sets of data. Since we are comparing the normalized CT responses to NaCl + Bz before and after FII_{im}, FII_m or FII_m sub-fractions in the same CT preparation, paired t-test was used to evaluate statistical significance.

Table 1. Taste stimuli used for CT experiments.

	(mM)	Stimuli	(mM)
R	10 KCl + 10 HEPES	NaCl	10 KCl + 10 HEPES + 100 NaCl
R	10 KCl + 10 HEPES	NaCl + Bz	10 KCl + 10 HEPES + 100 NaCl + 0.005 Bz
R	10 KCl + 10 HEPES	NaCl + Bz + RTX	10 KCl + 10 HEPES + 100 NaCl + 0.005 Bz + RTX (0–0.01)
R	10 KCl + 10 HEPES	NaCl + Bz + FII or sub-fractions	10 KCl + 10 HEPES + 100 NaCl + 0.005 Bz + FII or sub-fractions
R	10 KCl + 10 HEPES	NaCl + SB	10 KCl + 10 HEPES + 100 NaCl + 0.001 SB
R	10 KCl + 10 HEPES	NaCl + SB + FII	10 KCl + 10 HEPES + 100 NaCl + 0.001 SB + FII or sub-fractions)
R	10 KCl + 10 HEPES	N + Bz + SB + FII	10 KCl + 10 HEPES + 100 NaCl + 0.005 Bz + 0.001 SB + FII or sub-fractions
R	10 KCl	MSG + Bz + SB	10 KCl + 100 MSG + 0.005 Bz + 0.001 SB
R	10 KCl	MSG + Bz + SB + IMP	10 KCl + 100 MSG + 0.005 Bz + 0.001 SB + 1 IMP
R	10 KCl	MSG + Bz + SB + FII	10 KCl + 100 MSG + 0.005 Bz + 0.001 SB + FII
R	10 KCl	Control-1	300 NH_4Cl
R	10 KCl	Control-2	300 NaCl

4-(2-Hydroxyethyl)-1-piperazine ethanesulfonic acid (HEPES) was used to buffer the pH of rinse and salt stimuli at pH 6.1. Bz (benzamil); RTX (resiniferatoxin); SB (SB-366791, N-(3-methoxyphenyl)-4-chlorocinnamide). All compounds were obtained from Sigma-Aldrich.

2.3. Behavior Studies in Mice

Rats have a high preference for NaCl even in the presence of Bz [29]. Because of already high background NaCl preference, small increases in NaCl preference are difficult to evaluate in rats. In contrast, mice demonstrate a more moderate preference for NaCl and small shifts in the preference curve are easily detected. Therefore, mice were used for behavioral studies. Behavioral studies were performed in WT (C57BL/6J) mice (30–40 gm) using standard two bottle/48 h tests [39]. Both males and females were used. The care and use of the mice followed the institutional and national guidelines, and the protocol was approved by the committee on the Ethics of Animal Experiments of the Korea Food Research Institute (Permit Number: KFRI-M-12028). Mice (63–70 days of age) were housed in separated cages and were maintained on a standard laboratory chow (Pico-Lab Rodent Diet 20–5053, PMI Feeds) and water *ad libitum*. The air-conditioned animal room was maintained at 22 ± 2 °C, with relative humidity of 59 ± 1% and a 12 h light/dark cycle (light period, 07:00–19:00 h). Each mouse was tested at approximately the same time of day. Before the start of the experiment mice were given two bottles with water for 2 weeks. The experiment was started when mice were accustomed to drinking equally from 2 bottles. Mice were given a choice between two bottles, one containing water and the other a test solution in the following order: water, 30 mM NaCl, 80 mM NaCl, 100 mM NaCl, 120 mM NaCl, 150 mM NaCl, 200 mM NaCl and 300 mM NaCl. We also performed behavioral studies when both water and the NaCl solutions contained 10 μM amiloride. In some experiments, mice were given a choice between water and 100 mM NaCl or between water + 10 μM amiloride and 100 mM NaCl + 10 μM amiloride containing varying concentrations (0.1, 0.25, 0.5, 0.75, and 1.0%) of FII_m.

For each 48 h period the mass of water versus the mass of the test solution consumed by each mouse/g BW was measured. The preference ratio for a taste stimulus was calculated as the mass of the test solution consumed/48 h/g BW divided by the mass of the total fluid intake (mass water + mass of the test solution)/48 h/g BW). The bottles containing water or the test solution were switched from left to right every day. The data were analyzed using one-sample t-tests against 0.5, a reference value meaning indifference of the test solution with respect to the control solution.

2.4. Human Sensory Evaluation

All human sensory evaluation protocols were approved by the Public Institutional Review Board Designated by Ministry of Health and Welfare, South Korea. The ethic approval code is P01-202004-23-004. Each participant signed a consent form to participate in salt taste sensory evaluations. To maintain a subject's confidentiality, the personal data was coded and the taste data were analyzed off line. Previously trained panelists (men and women, ages between 25 to 37 years) with no history of basic taste disorders were recruited. The panelists washed their mouth after tasting each samples. The data was analyzed by one-way ANOVA to compare between-group differences.

2.4.1. Salt Sensory Evaluation

Panelists were trained to recognize salt taste intensity with reference to 0.2%, 0.35%, 0.5%, and 0.7% NaCl solution representing a value of 2.5, 5.0, 8.5, and 15.0, respectively, using a 15-point intensity scale [40]. To evaluate the effect of FII_{im} and FII_m on salt taste, FII_{im} or FII_m (0–0.01%) dissolved in 0.2% NaCl solution was presented to the panelists and the salt taste intensity was evaluated with reference to 0.2% NaCl (intensity scale value = 2.5; R1), and 0.35% NaCl (intensity scale value = 5.0; R2), respectively.

2.4.2. Umami Sensory Evaluation

According to the manufacturer's instructions, Japanese fish soup base, Hondashi (0.04 g) was dissolved in 100 ml water. The 0.04% Hondashi fish soup base was used as a control and was given an intensity of 5 on a 10-point intensity scale. FII_m at 0.003%, 0.01%, 0.03% and 0.3% was dissolved in 0.04% Hondashi Fish soup base and their effect was evaluated on umami taste by the same trained panelists ($n = 6$). As a control, FII_m was dissolved in water at 0.003%, 0.01%, 0.03% and 0.3%. These concentrations of FII_m were evaluated for umami taste by the same panelists ($n = 6$).

3. Results and Discussion

3.1. Effect of FII_m and FII_{im} on the Bz-insensitive NaCl CT Response

As shown in a representative CT trace (Figure 1A), adding increasing concentrations of FII_m to 100 mM NaCl + 5 μM Bz (NaCl + Bz) solution (Table 1) initially produced an increase in both phasic and tonic NaCl CT response of between 0.1% and 0.5%. Above 0.5% FII_m the magnitudes of the phasic and tonic CT responses were less than their respective maximum values. In the presence of 1% FII_m, the tonic CT response decreased below the NaCl + Bz CT response in the absence of FII_m (Figure 1A). The variation of the normalized mean tonic NaCl + Bz CT response plotted as a function of the log of FII_m or FII_{im} concentrations (%) are summarized in Figure 1B. FII_m produced a biphasic dose-response relationship for both the phasic (data not shown) and tonic (Figure 1B; •) NaCl + Bz CT response. The maximum increase in the mean normalized tonic CT response occurred at 0.5% of FII_m, an 88% increase relative to NaCl + Bz tonic CT response in the absence of FII_m. At 1% FII_m, the tonic NaCl + Bz CT response was significantly less than the tonic CT response with NaCl + Bz alone ($p = 0.0466$; $n = 3$). Stimulating the tongue with the rinse solution (R) containing varying concentrations of FII_m elicited only transient (phasic) CT responses that were concentration-independent and were indistinguishable from the mechanical rinse artifact (data not shown). These results indicate that, at the concentrations used in these experiments, FII_m, by itself, is not a gustatory stimulus in the fungiform taste receptive field and only modulate the CT response in the presence of salt (NaCl + Bz). In contrast, FII_{im} did not produce any changes in either the phasic (data not shown) or the tonic CT response between 0.1% and 1% (Figure 1B; ○). The Xyl-MRPs, GalA-MRPs, glucosamine-MRPs, and fructose-MRPs also produced biphasic effects on the Bz-insensitive NaCl CT response with maximum increase at 0.25%, 0.25%, 0.50%, 0.75% and 1%, respectively [27]. This suggests that both MRPs naturally generated during longer maturation and synthesized in vitro have a common property of producing biphasic effects on the Bz-insensitive NaCl CT response. The potency of MRPs depends upon the reacted sugar moiety. However, at present it is not known which sugar moieties are conjugated to the peptides comprising FII_m. FII_m is a mixture of MRPs of varying molecular weights, charge and affinity for their putative salt taste receptor(s). In comparison, 0.27% GalA-MRPs enhanced the Bz-insensitive NaCl CT response by 101% [27], suggesting naturally occurring FII_m produces effects on the NaCl + Bz CT response that are comparable to those produced by the GalA-MRPs.

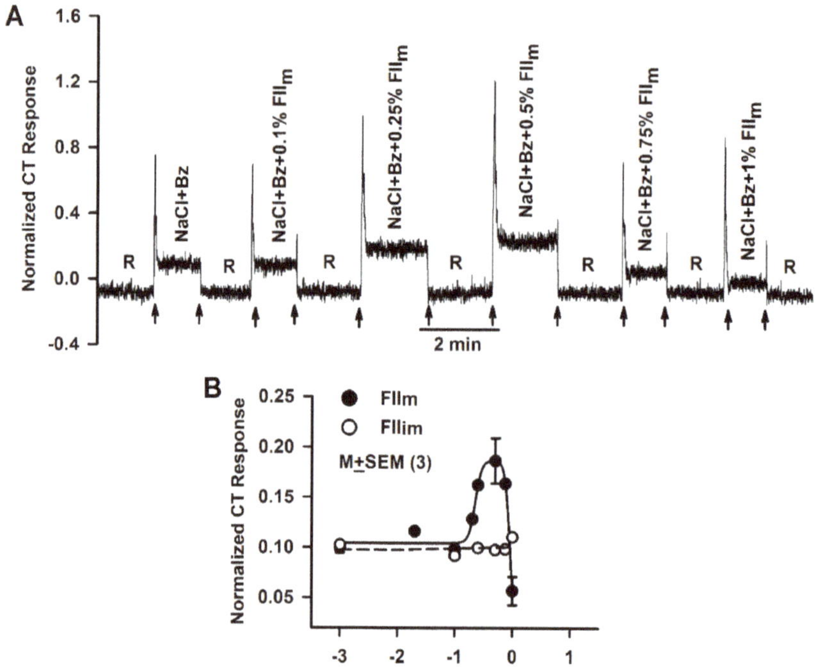

Figure 1. Effect of FII_{im} and FII_m on the benzamil (Bz)-insensitive NaCl chorda tympani (CT) response. (**A**) Shows a representative trace in which the CT responses were monitored while the rat tongue was first superfused with a rinse solution (R) and then with a stimulating solution containing 100 mM NaCl + 5 μM Bz + FII_m (0–1%) maintained at room temperature. The arrows represent the time periods when the rat tongue was superfused with R and the stimulating solutions. The data were normalized to the tonic response obtained with 0.3 M NH_4Cl. (**B**) Shows the mean normalized tonic NaCl CT responses in different sets of 3 rats each while their tongues were first stimulated with R and then with NaCl + Bz solutions containing 0–1% of the FII_m (•) or FII_{im} (○) expressed in log units. The values are M ± SEM of 3 rats.

3.2. *Effect of SB and FII_m on the Bz-insensitive NaCl CT Response*

In our previous studies, Bz-insensitive NaCl CT responses in rodents were inhibited by TRPV1 blockers. In addition, Bz-insensitive NaCl CT responses were not observed in TRPV1 knockout mice [33]. Accordingly, in the next series of experiments we tested if FII_m effects on salt responses were also sensitive to SB, a TRPV1 blocker. Because 0.4% and 0.5% FII_m give almost equivalent CT responses (Figure 1B), we used 0.4% FII_m in these experiments. In mixtures containing NaCl + Bz + SB, the constitutive NaCl + Bz tonic CT response was inhibited to the rinse baseline level (Figure 2A,C). Subsequently, stimulating the rat tongue with solutions containing NaCl + Bz + SB + 0.4% FII_m significantly inhibited the CT nerve response relative to NaCl + Bz + 0.4% FII_m (Figure 2A,C; ** $p = 0.0001$; $n = 3$). These results suggest that both the constitutive amiloride- and Bz-insensitive CT response and the subsequent FII_m-induced increase in the CT response are SB-sensitive.

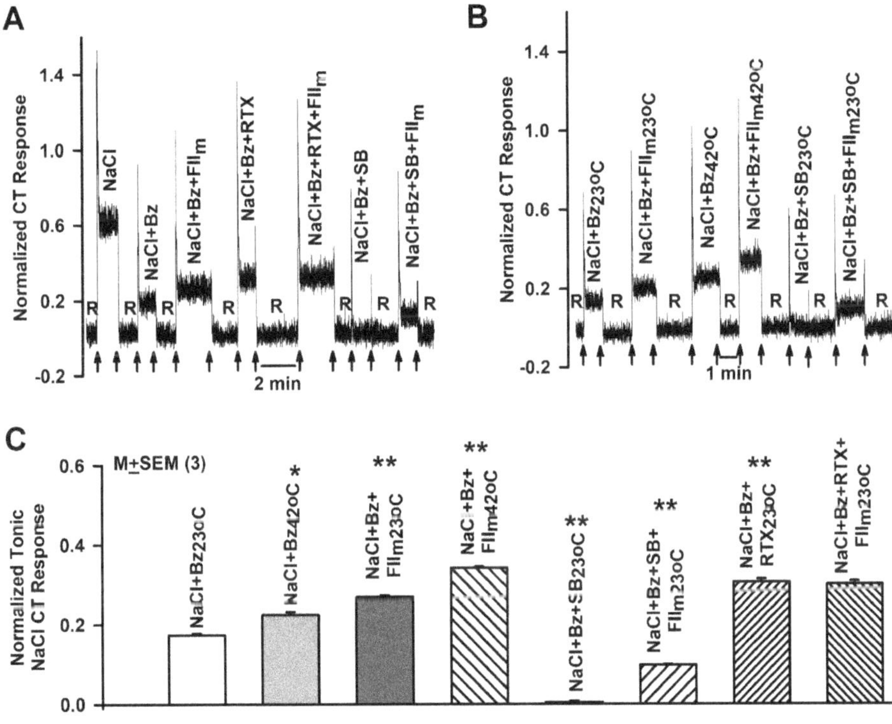

Figure 2. Effect of resiniferatoxin (RTX), SB-377791 (SB), FII_m and temperature on the benzamil (Bz)-insensitive NaCl chorda tympani (CT) response. (**A**) Shows a representative CT trace obtained while the rat tongue was first stimulated with rinse solution (R) and then with NaCl, NaCl + Bz, NaCl + Bz + 0.4% FII_m, NaCl + Bz + 0.25 μM RTX, NaCl + Bz + 0.4% FII_m + 0.25 μM RTX, NaCl + Bz + 1 μM SB and NaCl + Bz +SB + 0.4% FII_m maintained at room temperature. The data were normalized to the tonic response obtained with 0.3 M NH_4Cl. The arrows represent the time periods when the rat tongues were superfused with R and the stimulating solutions. (**B**) Shows a representative CT trace obtained while the rat tongue was first stimulated with R at 23 °C ($R_{23\,°C}$) and then with NaCl + Bz (NaCl + $Bz_{23\,°C}$), NaCl + Bz + 0.4% FII_m at 23 °C (NaCl + Bz + $FII_{m23\,°C}$), NaCl + Bz at 42 °C (NaCl + $Bz_{42\,°C}$) and NaCl + Bz + 0.4% FII_m at 42 °C (NaCl + Bz + $FII_{m42\,°C}$). The trace also shows the CT response in the presence of NaCl + Bz + SB and NaCl + Bz + SB + 0.4% FII_m maintained at 23 °C. The data were normalized to the tonic response obtained with 0.3 M NH_4Cl. The arrows represent the time periods when the rat tongues were superfused with R and the stimulating solutions. (**C**) Shows the M ± SEM normalized rat tonic NaCl + Bz CT responses at 23 °C and 42 °C in the absence and presence of 0.4% FII_m. All unpaired comparisons were made with respect to the normalized value of the tonic CT response to NaCl + Bz at 23 °C. * $p = 0.0038$; ** $p = 0.0001$; $n = 3$).

3.3. Effect of RTX and FII_m on the NaCl + Bz CT Response

Consistent with previous studies [33,35], at room temperature (23 °C), RTX (0.25 μM) enhanced the rat NaCl + Bz CT response relative to NaCl + Bz (Figure 2A,C; * $p = 0.0001$; $n = 3$). When the tongue was stimulated with NaCl + Bz solutions containing both RTX (0.25 μM) and FII_m (0.4%), no further increase in the magnitude of the Bz-insensitive NaCl CT response was observed relative to NaCl + Bz + RTX (Figure 2C). These results suggest that RTX and FII_m target the same amiloride-insensitive pathway(s). The Bz-insensitive NaCl taste responses are regulated by several intracellular signaling mediators. A decrease in taste cell Ca^{2+}, activation of protein kinase C, and inhibition of calcineurin enhanced the magnitudes of the Bz-insensitive NaCl CT responses in the presence of RTX, and either minimized or completely eliminated the decrease in the CT response at RTX concentrations >1 μM. In contrast, increasing taste cell Ca^{2+} inhibited the Bz-insensitive NaCl CT response in the presence of RTX [41]. An increase in taste cell phosphatidylinositol 4,5-bisphosphate inhibited the control NaCl + Bz CT response and decreased its sensitivity to RTX. Alternately, a decrease in phosphatidylinositol 4,5-bisphosphate enhanced the control NaCl + Bz CT response, increased its sensitivity to RTX stimulation, and inhibited the desensitization of the CT response at RTX concentrations >1 μM [42]. It is likely that Bz-insensitive NaCl CT responses in the presence of FII_m are also regulated by the above intracellular modulators and are responsible for their biphasic effects on the NaCl CT response.

3.4. Effect of Elevated Temperature and FII_m on the NaCl + Bz CT Response

In our previous studies, Bz-insensitive NaCl CT responses in rodents were temperature dependent. In addition, temperature and modulators of the Bz-insensitive NaCl CT response produced additive effects on CT response [26,33,36]. Accordingly, we next tested the effect of elevating the temperature from 23 °C to 42 °C on the CT response to NaCl + Bz and NaCl + Bz + 0.4% FII_m. As shown in a representative CT recording, elevating the temperature from 23 °C to 42 °C increased the magnitude of the tonic NaCl + Bz CT response relative to 23 °C (Figure 2B). FII_m (0.4%) further increased the CT response at 23 °C and 42 °C (Figure 2B). The mean tonic NaCl + Bz CT response at 23 °C (Figure 2C) was significantly enhanced by increasing the temperature to 42 °C (* $p = 0.0039$) and by the addition of 0.4% FII_m (Figure 2C; ** $p = 0.0001$; $n = 3$). These results show that elevated temperature and FII_m produce additive effects on the amiloride-insensitive NaCl CT response.

3.5. Effect of FII_m sub-fractions of Different Molecular Weights ($FII_{m(a\text{-}d)}$) on the NaCl + Bz CT Response

FII_m was further separated into four sub-fractions of varying molecular weights: FII_{ma} (500–1000 Da), FII_{mb} (1000–3000 Da), FII_{mc} (3000–5000 Da) and FII_{md} (5000–10,000 Da). As shown in representative CT recordings in Figs. 3A and 3B, the relationship between varying concentrations of FII_{ma} and FII_{mc} and the magnitude of NaCl + Bz CT response was shifted to the right on the concentration axis relative to FII_m (Figure 1A). The relationships between varying concentrations of FII_{ma}, FII_{mb}, FII_{mc} and FII_{md} and the corresponding mean normalized tonic NaCl + Bz CT response are plotted in Figure 3C. The results show that for all sub-fractions $FII_{m(a–d)}$, the relationship between their concentrations and the magnitude of tonic NaCl + Bz CT response is shifted to the right on the concentration axis relative to FII_m. FII_{ma} produced the maximum increase in the NaCl + Bz tonic CT response at a concentration between 1.5 and 2.5% (Figure 3C; ◆). This concentration is significantly higher than the concentration at which FII_m produced the maximum increase in the NaCl + Bz CT response (0.5%; Figure 1B). These results suggest that FII_m fraction is composed of MRPs of varying molecular weights that differ in their affinity and potency in modulating the putative amiloride-insensitive salt taste pathway(s).

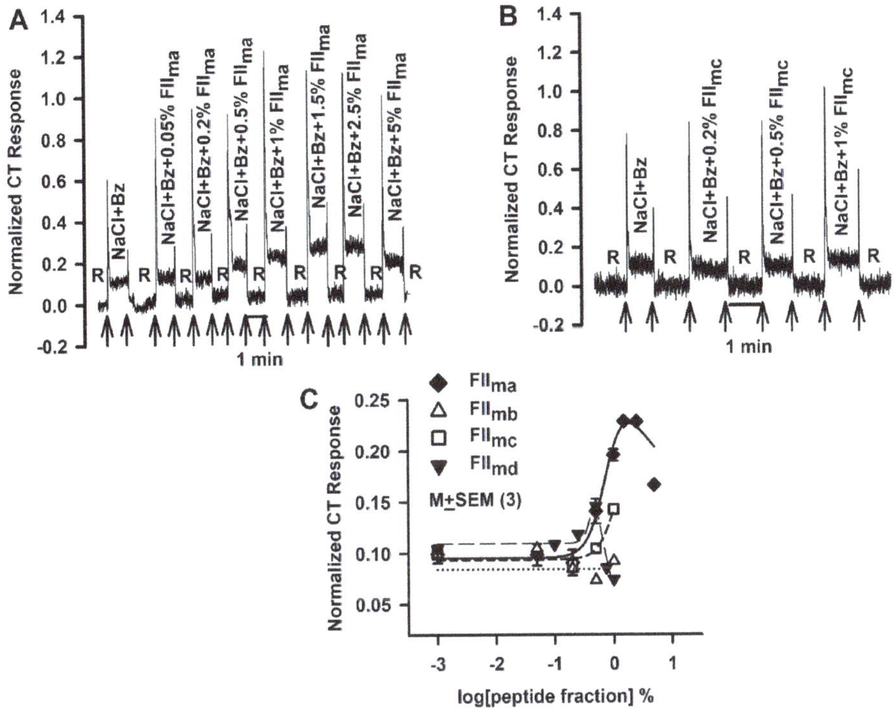

Figure 3. Effects of FII_m sub-fractions ($FII_{m(a-d)}$) on the benzamil (Bz)-insensitive NaCl chorda tympani (CT) response. Representative CT responses showing the effect of adding varying concentrations of FII_m sub-fractions FII_{ma} (500–1000 Da) (**A**) and FII_{mc} (1000–3000 Da) (**B**) on the rat CT responses to NaCl + Bz. The arrows represent the time period when the tongue was superfused with the rinse and stimulating solutions. In each rat the data were normalized to the tonic response obtained with 0.3M NH_4Cl. (**C**) Shows the mean normalized tonic NaCl CT responses in different sets of 3 rats each while their tongues were first stimulated with R and then with NaCl + Bz solutions containing 0–1% of the four FII_m sub-fractions in log units. The values are M ± SEM of 3 rats in each group. In each case the data were fitted to Equation (4).

3.6. *Effect of FII_m sub-fractions (Neutral, Acidic, Basic and Aromatic) on the NaCl + Bz CT Response*

FII_m was further separated into neutral, acidic, basic and aromatic sub-fractions. Since the relationships between varying concentrations of the neutral, acidic and basic fractions and the magnitude of the tonic NaCl + Bz CT response were very similar in individual rats, the data from these three fractions were combined and are plotted in Figure 4A (▲). In all three fractions, the relationship between their concentrations and the magnitude of the mean normalized tonic NaCl CT response was shifted to the right on the concentration axis relative to FII_m (Figure 4A; •). In contrast, the aromatic fraction produced a biphasic response in the NaCl + Bz CT response with a very sharp-peak at 0.75% (Figure 4A; ○). These results further suggest that FII_m is composed of neutral, acidic, basic and aromatic MRPs that show varying degrees of potency and affinity for modulating the putative amiloride-insensitive salt taste pathway(s). It is interesting to note that relative to control (NaCl + Bz), 0.5% FII_m (Figure 1A) produced an equivalent maximum increase in the tonic NaCl + Bz CT response as 1 μM RTX (Figure 4B).

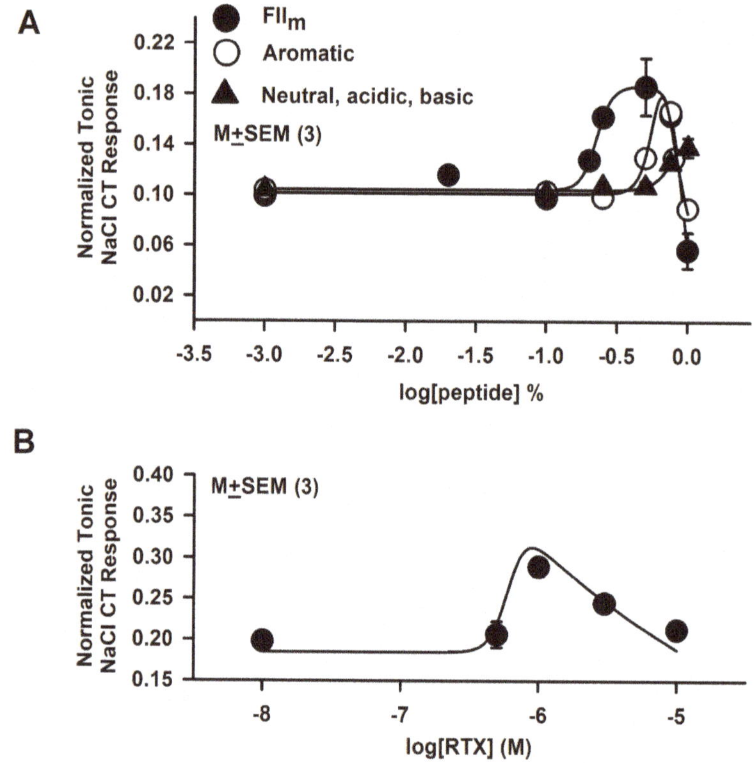

Figure 4. Effects of aromatic, neutral, acidic and basic FII_m sub-fractions on the benzamil-insensitive NaCl chorda tympani (CT) response. (**A**) Shows the relationship between varying FII_m sub-fraction concentrations expressed in log units and the mean normalized tonic NaCl CT response from 3 rats in each group for FII_m (•), aromatic (○) and combined neutral, acidic and basic maillard reacted peptides (▲). (**B**) Shows the relationship between resiniferatoxin (RTX) concentrations expressed in log units and the mean normalized tonic NaCl CT responses from 3 rats (•). The values are M ± SEM of 3 rats in each group.

We also recorded FII_m-induced changes in the Bz-insensitive NaCl CT response in wild type (WT; C57BL/6J) and homozygous TRPV1 knockout mice (B6. 129S4-Trpv1^{tmijul}; Jackson Laboratory, Bar Harbor, ME). Consistent with our earlier study with MRPs [27], FII_m produced a similar biphasic response on the Bz-insensitive NaCl CT response in WT mice. In TRPV1 KO mice, FII_m (0.4%) did not induce CT response above the rinse baseline (data not shown). This is akin to our results in rats. SB inhibited the basal Bz-insensitive NaCl CT response. In the continuous presence of SB, FII_m produced a significantly smaller increase in the Bz-insensitive NaCl CT response relative to the absence of SB (Figure 2). These results indicate that FII_m produces similar effects on rats and mice.

3.7. *Effect of Calcitonin Gene Related Peptide (CGRP) on NaCl CT Responses*

RTX activates and SB inhibits amiloride-insensitive NaCl CT responses (Figure 2). However, TRPV1 immunoreactivity was not found in TRCs [43–45]. We hypothesize that RTX and other modulators of TRPV1 alter Bz-insensitive NaCl CT responses indirectly, by releasing CGRP from trigeminal nerves [46]. The released CGRP then acts on its specific receptor (CGRPR) in TRCs to modulate Bz-insensitive NaCl CT responses [47].

Due to the concern that topical lingual application of CGRP, a large neuropeptide, may not be able to reach its receptor in TRCs, CGRP was administered by intraperitoneal injection. CT responses were monitored while the rat tongue was stimulated with 0.3M NH_4Cl, 0.3 M NaCl and 0.1M NaCl before and after an i.p. injection of 23 μg/100 BW or 68 μg/100 g BW CGRP dissolved in 0.5 mL PBS. Following i.p. injection of 68 μg/100 g BW CGRP the NaCl CT response increased with time (data not shown). As shown in Figure 5A, 10 min post CGRP injection, the NaCl CT responses were almost two times greater than control (Figure 5B). CGRP also induced an increase in the CT response to 0.3M NaCl. However, an i.p. injection of 23 μg/100 BW CGRP did not induce any changes in rat NaCl CT response 10 min post CGRP injection (Figure 5B). These results indicate that CGRP effects on NaCl CT response are both time- and dose-dependent and are observed over a range of NaCl concentrations. These results suggest a possible interaction between the trigeminal and salt taste systems.

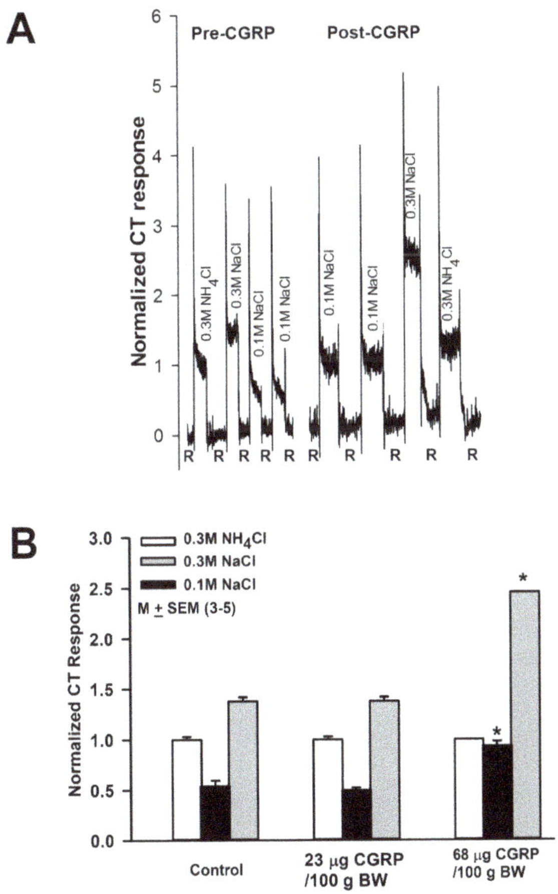

Figure 5. Effect of i.p. injection of calcitonin gene related peptide (CGRP) on NaCl chorda tympani (CT) response. (**A**) Shows a representative CT trace obtained while the rat tongue was first stimulated with rinse solution (R) and then with 0.3M NH_4Cl, 0.3M NaCl and 0.1M NaCl before and after i.p. injection of CGRP (68 μg/100 g BW in PBS). In each rat the data were normalized to the tonic response obtained with 0.3M NH_4Cl. The values are M ± SEM of 3 rats in each group. (**B**) Shows summary of the data from 3 rats in each group injected with either 23 or 68 μg CGRP/100 g BW. Values are M ± SEM of 3 rats. * p = 0.017 (0.1M NaCl) and 0.009 (0.3M NaCl).

Using calcium imaging, a subset of acid responsive Type III mouse circumvallate TRCs were identified as the amiloride-insensitive salt responsive cells [2]. CGRPR has been suggested as the functional link to cellular transduction pathway for CGRP action on Type III TRCs. CGRP has been shown to increase [Ca^{2+}] in Type III TRCs. This effect of CGRP was dependent upon phospholipase C activation and was prevented by U73122 [47]. In mouse taste buds, CGRP caused TRCs to secrete serotonin (5-HT), a presynaptic (Type III) cell transmitter. 5-HT seems to reduce taste evoked ATP secretion in Type II cells [47]. However, at present this information is lacking in the fungiform taste receptive field.

Here, we present new data that suggest that CGRP can modulate rat amiloride-insensitive NaCl CT responses (Figure 5). In a recent study [3] amiloride-insensitive Ca^{2+} responses in mouse taste bud cells were localized to the apical tips of Type II, but not in Type III fungiform TRCs. It is suggested that, because anterior (fungiform) and posterior (circumvallate) taste fields differ functionally, in an earlier study [2] amiloride-insensitive NaCl responses may have been detected in only Type III circumvallate taste cells. Although the identity and location of the amiloride- and Bz-insensitive pathway(s) are ambiguous at present, CT recordings demonstrate that a component of the amiloride- and Bz-insensitive NaCl CT response at low NaCl concentrations (100 mM) is present in the anterior taste field that is modulated by RTX, FII_m, temperature, SB (Figures 1A and 2B) and voltage [33,34].

N-geranyl cyclopropylcarboxamide (NGCC), a modulator of the amiloride- and Bz-insensitive NaCl CT responses [48], activates hTRPV1 expressed in HEK293T cells [49]. In our preliminary studies, component of FII_m induced inward current in TRPV1-expressing cells in whole-cell patch-clamp recordings [50,51]. Currently studies are underway to demonstrate direct activation of the expressed umami taste receptor by FII_m. However, at present it is not clear if, like RTX, other modulators of the amiloride-insensitive pathway release CGRP from TRPV1 expressed in trigeminal neurons in a dose-dependent manner. In addition, it is also not known if, like other modulators of the amiloride-insensitive NaCl CT response, CGRP elicits a biphasic effect on rat NaCl CT responses. Taken together, our data suggest a possible linkage between the trigeminal system and amiloride-insensitive salt taste. It has recently been demonstrated that sour taste pathway works together with the somatosensory system to trigger aversive responses to acidic stimuli [5].

3.8. Behavioral Studies with Mice

Under control conditions, mice demonstrated a bell shaped NaCl preference curve with a significant preference for 100 mM NaCl (Figure 6A; ○; * $p = 0.02$; $n = 10$) and aversion for 300 mM NaCl (**** $p = 0.0001$) [39]. In the presence of 10 μM amiloride the NaCl preference curve was again biphasic but was shifted to the right on the NaCl concentration axis. In the presence of amiloride, mice showed a significant NaCl preference at 150 mM NaCl (Figure 6 A; •; *** $p = 0.0024$).

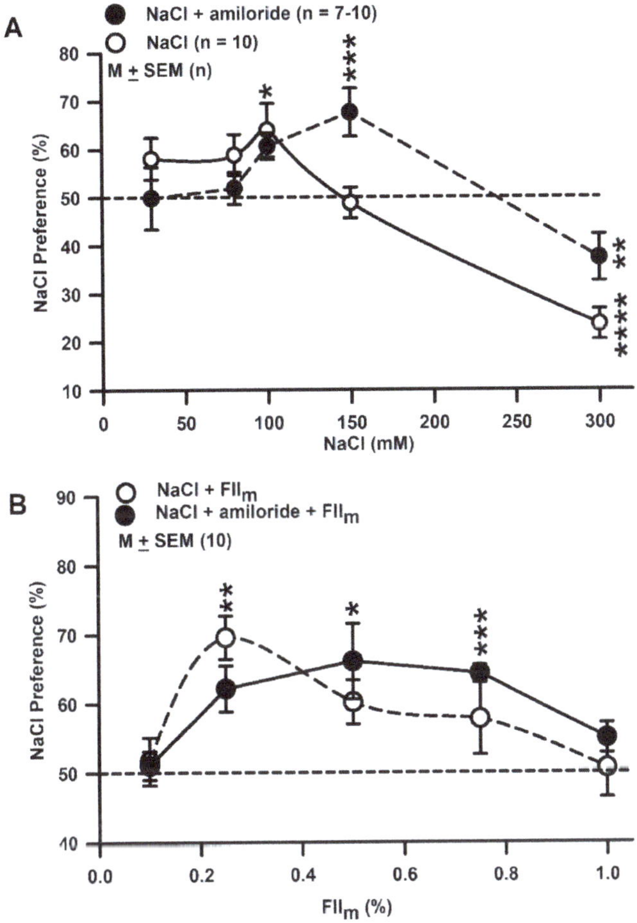

Figure 6. Effect of amiloride and FII$_m$ on NaCl Preference in WT mice. (**A**) Shows NaCl Preference in WT mice when given a choice between H_2O and varying concentrations of NaCl (3, 80, 100, 120, 150, 200 and 300 mM) in the absence (○) and presence of 10 µM amiloride (•). The values are presented as mean (M) ± SEM of *n*, where *n* = 7–10. * *p* = 0.02; ** *p* = 0.0134; *** *p* = 0.0024; **** *p* = 0.0001. (**B**) Shows NaCl Preference in WT mice when given a choice between H_2O and 100 mM NaCl (○) or H_2O and 100 mM NaCl + 10 µM amiloride (•) containing increasing concentrations of FII$_m$ (0.1 to 1%). * *p* = 0.0086; ** *p* = 0.0018; *** *p* = 0.0001 (*n* = 10). Dotted line represents the indifference value.

As shown in Figure 6B, adding increasing concentrations of FII$_m$ (0.1 to 1%) to 100 mM NaCl solutions in the absence and presence of 10 µM amiloride produced biphasic changes in NaCl preference, increasing it at 0.25% and lowering it at higher concentrations. Under control conditions, FII$_m$ maximally enhanced the NaCl preference at 0.25% relative to NaCl alone (Figure 6B; ○; ** *p* = 0.0001; *n* = 10). Above 0.25% FII$_m$ NaCl preference was significantly less than its maximum value. In the presence of 10 µM amiloride the maximum increase in NaCl preference was observed at 0.5% FII$_m$ (Figure 6B; •; * *p* = 0.0086). Above 0.5% FII$_m$ NaCl preference was significantly less than its maximum value. There was no change in NaCl preference when equivalent concentrations of the FII$_{im}$ were added to the test solutions containing 100 mM NaCl or 100 mM NaCl + 10 µM amiloride (data not shown). These behavioral responses to NaCl are correlated with the biphasic effects of FII$_m$ concentrations on the amiloride- and Bz-insensitive NaCl CT responses (Figure 1). In this sense, FII_m mimics the effect of other modulators [27,33,34,36,48,52] of the amiloride- and Bz-insensitive NaCl CT responses.

3.9. Effect of FII_m on Salt Taste in Human

In human sensory evaluation, FII_m produced a biphasic effect on salt taste. FII_m increased salt taste intensity between 0.3 and 0.5%, but slightly decreased it above 0.5% (Figure 7). The maximum salt taste intensity in human subjects was detected at 0.5% FII_m (Figure 7; •). In contrast, FII_{im} had no significant effect on human salt taste perception (Figure 7; ○). In our previous studies, GalA-MRPs, Xyl-MRPs [27] and NGCC [48], modulators of the amiloride- and Bz-insensitive NaCl CT responses in rodents, also produced biphasic effects on human salt taste intensity. Although functional ENaC channels are expressed in human fungiform TRCs [53], the amiloride- and Bz-insensitive salt taste receptors are the predominant transducers of salt taste in humans [25–27]. Thus, modulation of the amiloride-insensitive salt taste in humans via FII_m or other modulators may provide alternate ways to regulated human salt taste and perhaps salt intake.

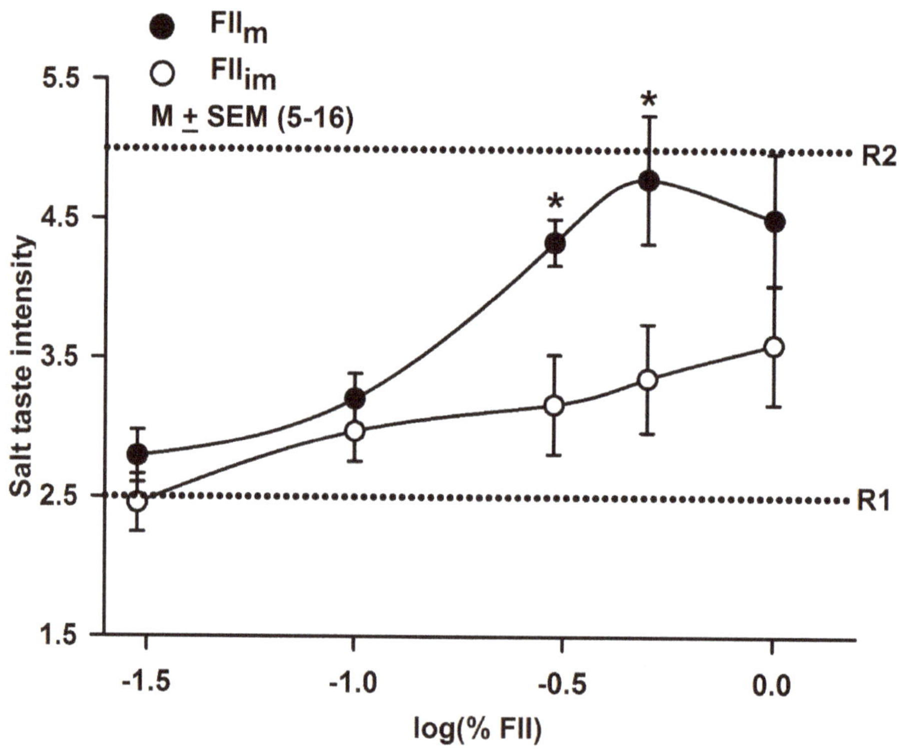

Figure 7. Effect of FII_m and FII_{im} on human salt taste intensity. Shows the effect of varying concentrations (0.03 and 1.0%) of FII_{im} (○) and FII_m (•) expressed in log units on human salt taste intensity. R1 corresponds to the intensity (2.5) of 0.2% NaCl and R2 corresponds to the intensity (5.0) of 0.35% NaCl. FII_m showed a significant (* $p = 0.01$) salt taste-enhancing activity at 0.003% and 0.005%. In contrast, no effect of FII_{im} was observed on human salt taste intensity over the concentration range between 0.03 and 1.0%.

3.10. Effect of FII_m on the Rat CT Response to Glutamate

Stimulating the rat tongue with 100 mM MSG + Bz + SB elicited a CT response and the CT response was enhanced in the presence of 1 mM IMP (Figure 8A). Glutamate CT response was also enhanced in the presence of 2.5% FII_m. The normalized tonic CT responses to glutamate in the absence and presence of IMP and FII_m are summarized in Figure 8B. FII_m at 2.5% enhanced the CT response to glutamate that was equivalent to the enhancement observed with 1 mM IMP. These results further suggest that unlike the Bz-insensitive NaCl CT response, the basal umami CT response and the subsequent FII_m induced enhancement of the umami CT response is SB-insensitive. In our previous study, Xyl-MRs also enhanced the CT response to glutamate at concentrations above which they modulated the NaCl + Bz CT responses. In contrast to Xyl-MRPs, at these concentrations GalA-MRPs or Glc-NH_2-MRPs did not show effects on the glutamate CT response [27]. These results suggest that the umami enhancing effect of MRPs is dependent on the conjugated sugar(s). However, at present the identity of the specific sugar resides conjugated with the peptides comprising the FII_m is not known.

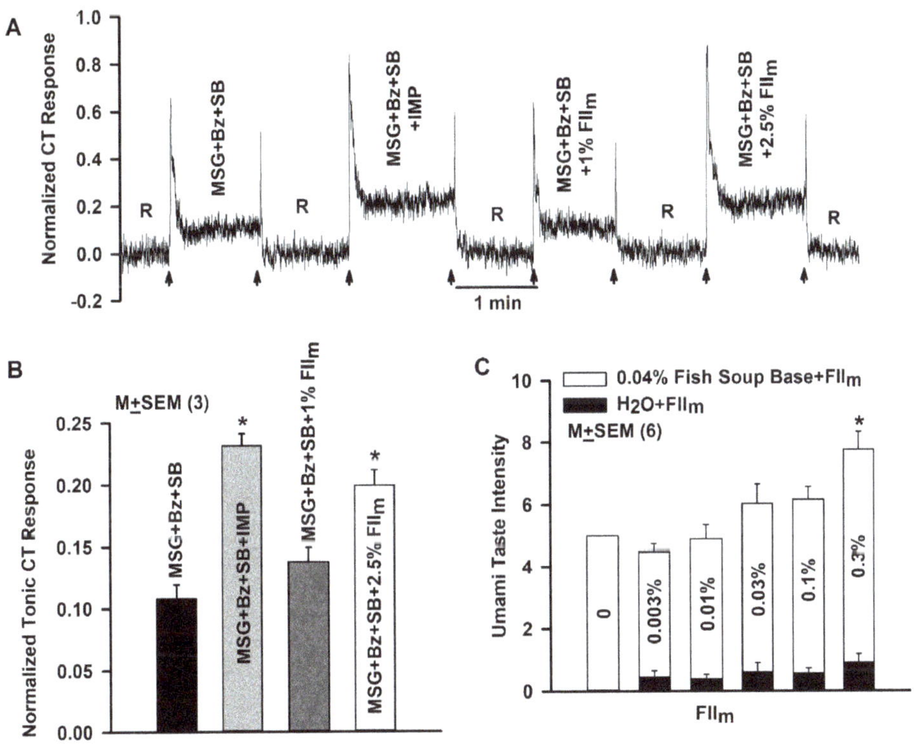

Figure 8. Effect of FII_m on the glutamate chorda tympani (CT) response and human umami taste sensory evaluation. (**A**) Shows a representative CT response in which the rat tongue was first rinsed with the rinse solution (R) and then with 100 mM MSG + 5 μM benzamil (Bz) + 1 μM SB-366791 (SB), MSG + Bz + SB +1 mM IMP, MSG + Bz + SB + 1% FII_m and MSG + Bz + SB + 2.5% FII_m. The arrows represent the time period when the tongue was superfused with the rinse and stimulating solutions. (**B**) Shows mean normalized tonic CT responses from 3 rats. In each rat the data were normalized to the tonic response obtained with 0.3 M NH_4Cl. * $p = 0.001$. (**C**) Shows the effect of adding increasing concentrations of FII_m (0.003 to 0.3%) to the 0.04% Fish Soup Base (open bars) or to H_2O (filled bars). The values are presented as M ± SEM of n, where n represents the number of panel members tested. * $p = 0.01$.

In our earlier study [33], RTX demonstrated a biphasic response on the rat Bz-insensitive NaCl CT response. At 1 μM, it maximally enhanced and at 10 μM, maximally inhibited the Bz-insensitive NaCl CT response. At 1 and 10 μM concentrations, RTX did not alter CT responses to 500 mM sucrose,

10 mM quinine and 10 mM HCl. These results tend to suggest that over the concentration range that alter the Bz- insensitive NaCl CT response, modulators of the amiloride-insensitive pathway may not alter sweet, bitter or sour taste. At present it is not known if FII_m concentrations that modulate salt responses also alter responses to other taste stimuli.

3.11. Effect of FII_m on Umami Taste in Human

In human subjects, adding 0.3% FII_m to umami soup base produced a significant increase in umami taste intensity (Figure 8C; open bars; $p < 0.01$; $n = 9$). Equivalent concentrations of FII_m added to water produced umami intensity ratings of < 1 (Figure 8C; filled bars). In contrast to a strong salt taste enhancing effect at 0.5% FII_m, lower concentrations (<0.3%) of the FII_m did not have a significant effect on umami taste intensity. Thus, depending upon the concentration, MRPs can be used either as salt taste or umami taste modifiers.

These results show that FII_m modulates both salt and umami taste in humans but at different concentration range. The differences in the sensitivity to FII_m between humans and mice are most likely due to the variations in the umami taste receptor protein [54,55]. This dual property of being able to modulate the Bz-insensitive NaCl response and the glutamate response at two different concentration ranges is not restricted to FII_m. We have recently shown that at different concentrations ranges, NGCC modulates Bz-insensitive salt taste responses and glutamate taste responses in humans and animal models [48].

4. Conclusions

In summary, a naturally occurring kokumi taste active peptide fraction (MW 500–10,000 Da) isolated from mature (FII_m; 4-year old) Ganjang, a typical Korean Soy Sauce, modulates the amiloride-, Bz-insensitive NaCl CT response in rodents in a biphasic manner. At low concentrations (0.1 to 0.5%) it enhanced and at higher concentrations (>0.5%) inhibited the Bz-insensitive NaCl CT response. FII_m effects on Bz-insensitive NaCl CT responses are TRPV1 dependent. FII_m may indirectly alter CT responses to NaCl via the release of CGRP from trigeminal fibers near the fungiform taste buds in the anterior taste field. This suggests a novel relationship between trigeminal system and salt taste perception. At concentrations >1%, FII_m enhanced the CT response to glutamate. In human sensory tests, FII_m increased the salt taste intensity between 0.3 and 0.5%, and the umami taste intensity at 0.3%. We conclude that, depending upon its concentration, FII_m modulates both salty and umami tastes in humans and rodents. The active component(s) and salt enhancing property of naturally occurring MRPs by longer maturation in food should be further investigated for a better understanding of the potential link between the compound and its beneficial effect in reducing salt intake in the human population.

Author Contributions: Conceptualization, M.-R.R. and V.L.; methodology, E.-Y.K., S.M.; validation, E.-Y.K.; formal analysis, A.-Y.S., H.-J.S., J.Q; investigation, A.-Y.S., H.-J.S., S.M., J.Q.; data curation, Y.K.; writing—original draft preparation, M.-R.R. and V.L.; writing—review and editing, M.-R.R. and V.L..; funding acquisition, M.-R.R., V.L. and J.R.G. All authors have read and agreed to the published version of the manuscript.

Funding: This research was funded by Korea Food Research Institute, grant numbers E069002 and E0131201, the National Research Foundation of Korea (NRF), grant number NRF2020R1A2C2004661, National Institute of Health, grant numbers DK34153 (J.R.G.), DC-005981 (V.L.), and DC-011569 (V.L.), VCU Wright Center for Clinical and Translational Research 2019 and VCU Health under the Value and Efficiency Teaching and Research program 2019 (V.L.). The APC was funded by NRF.

Acknowledgments: We gratefully acknowledge the assistance of MJ Kim of Korea Food Research Institute.

Conflicts of Interest: The authors declare no conflict of interest. The funders had no role in the design of the study; in the collection, analyses, or interpretation of data; in the writing of the manuscript, or in the decision to publish the results.

References

1. Yarmolinsky, D.A.; Zuker, C.S.; Ryba, N.J. Common sense about taste: From mammals to insects. *Cell* **2009**, *139*, 234–244. [CrossRef] [PubMed]
2. Lewandowski, B.C.; Sukumaran, S.K.; Margolskee, R.F.; Bachmanov, A.A. Amiloride-Insensitive Salt Taste Is Mediated by Two Populations of Type III Taste Cells with Distinct Transduction Mechanisms. *J. Neurosci.* **2016**, *36*, 1942–1953. [CrossRef] [PubMed]
3. Roebber, J.K.; Roper, S.D.; Chaudhari, N. The Role of the Anion in Salt (NaCl) Detection by Mouse Taste Buds. *J. Neurosci.* **2019**, *39*, 6224–6232. [CrossRef] [PubMed]
4. Nomura, K.; Nakanishi, M.; Ishidate, F.; Iwata, K.; Taruno, A. All-Electrical Ca^{2+}-Independent Signal Transduction Mediates Attractive Sodium Taste in Taste Buds. *Neuron* **2020**, *106*, 1–14. [CrossRef]
5. Zhang, J.; Jin, H.; Zhang, W.; Ding, C.; O'Keeffe, S.; Ye, M.; Zuker, C.S. Sour Sensing from the Tongue to the Brain. *Cell* **2019**, *179*, 392–402. [CrossRef]
6. Teng, B.; Wilson, C.E.; Tu, Y.H.; Joshi, N.R.; Kinnamon, S.C.; Liman, E.R. Cellular and Neural Responses to Sour Stimuli Require the Proton Channel Otop1. *Curr. Biol.* **2019**, *29*, 3647–3656. [CrossRef]
7. Keast, S.J.R.; Breslin, P.A.S. An overview of binary taste-taste interactions. *Food Qual. Prefer.* **2003**, *14*, 111–124. [CrossRef]
8. Roper, S.D.; Chaudhari, N. Taste buds: Cells, signals and synapses. *Nat. Rev. Neurosci.* **2017**, *18*, 485–497. [CrossRef]
9. Yamaguchi, S. Basic properties of umami and effects on humans. *Physiol. Behav.* **1991**, *49*, 833–841. [CrossRef]
10. Zhang, F.; Klebansky, B.; Fine, R.M.; Xu, H.; Pronin, A.; Liu, H.; Tachdjian, C.; Li, X. Molecular mechanism for the umami taste synergism. *Proc. Natl. Acad. Sci. USA* **2008**, *105*, 20930–20934. [CrossRef]
11. Kim, M.J.; Son, H.J.; Kim, Y.; Misaka, T.; Rhyu, M.R. Umami-bitter interactions: The suppression of bitterness by umami peptides via human bitter taste receptor. *Biochem. Biophys. Res. Commun.* **2015**, *456*, 586–590. [CrossRef]
12. Servant, G.; Tachdjian, C.; Tang, X.Q.; Werner, S.; Zhang, F.; Li, X.; Kamdar, P.; Petrovic, G.; Ditschun, T.; Java, A.; et al. Positive allosteric modulators of the human sweet taste receptor enhance sweet taste. *Proc. Natl. Acad. Sci. USA* **2010**, *107*, 4746–4751. [CrossRef] [PubMed]
13. DuBois, G.E.; Prakash, I. Non-caloric sweeteners, sweetness modulators, and sweetener enhancers. *Annu. Rev. Food Sci. Technol.* **2012**, *3*, 353–380. [CrossRef] [PubMed]
14. Ueda, Y.; Sakaguchi, M.; Hirayama, K.; Miyajima, R.; Kimizuka, A. Characteristic Flavor Constituents in Water Extract of Garlic. *Agric. Biol. Chem.* **1990**, *54*, 163–169.
15. Maruyama, Y.; Yasuda, R.; Kuroda, M.; Eto, Y. Kokumi substances, enhancers of basic tastes, induce responses in calcium-sensing receptor expressing taste cells. *PLoS ONE* **2012**, *7*, e34489. [CrossRef] [PubMed]
16. Ueda, Y.; Tsubuku, T.; Miyajima, R. Composition of Sulfur-Containing Components in Onion and Their Flavor Characters. *Biosci. Biotechnol. Biochem.* **1994**, *58*, 108–110. [CrossRef] [PubMed]
17. Ueda, Y.; Yonemitsu, M.; Tsubuku, T.; Sakaguchi, M.; Miyajima, R. Flavor characteristics of glutathione in raw and cooked foodstuffs. *Biosci. Biotechnol. Biochem.* **1997**, *61*, 1977–1980. [CrossRef]
18. Wang, M.; Yao, Y.; Kuang, D.; Hampson, D.R. Activation of family C G-protein-coupled receptors by the tripeptide glutathione. *J. Biol. Chem.* **2006**, *281*, 8864–8870. [CrossRef]
19. Ohsu, T.; Amino, Y.; Nagasaki, H.; Yamanaka, T.; Takeshita, S.; Hatanaka, T.; Maruyama, Y.; Miyamura, N.; Eto, Y. Involvement of the calcium-sensing receptor in human taste perception. *J. Biol. Chem.* **2010**, *285*, 1016–1022. [CrossRef]
20. Maillard, L.C. Réaction de Maillard. Action des acides aminés sur les sucres: Formation des mélanoïdines par voie méthodique. *Compte-Rendu De L'académie Des. Sci.* **1912**, *154*, 66–68.
21. Rhyu, M.R.; Kim, E.Y.; Ogasawara, M.; Nakahisaki, H. Maillard peptides are a clue to the Kokumi taste of a typical Korean soy sauce, Ganjang. In Proceedings of the Association for Chemoreception Sciences, Sarasota, FL, USA, 13–17 April 2005.
22. DeSimone, J.; Lyall, V. Amiloride-Sensitive Ion Channels. In *The Senses: A Comprehensive Reference*; Basbaum, A., Kaneko, A., Shepherd, G., Westheimer, G., Stuart, F., Gary, B., Eds.; Olfaction & Taste; Academic Press: San Diego, CA, USA, 2008; Volume 4, pp. 281–288.
23. Chandrashekar, J.; Kuhn, C.; Oka, Y.; Yarmolinsky, D.A.; Hummler, E.; Ryba, N.J.; Zuker, C.S. The cells and peripheral representation of sodium taste in mice. *Nature* **2010**, *464*, 297–301. [CrossRef] [PubMed]

24. Lin, W.; Finger, T.E.; Rossier, B.C.; Kinnamon, S.C. Epithelial Na+ channel subunits in rat taste cells: Localization and regulation by aldosterone. *J. Comp. Neurol.* **1999**, *405*, 406–420. [CrossRef]
25. Ossebaard, C.A.; Smith, D.V. Effect of amiloride on the taste of NaCl, Na-gluconate and KCl in humans: Implications for Na+ receptor mechanisms. *Chem. Senses* **1995**, *20*, 37–46. [CrossRef] [PubMed]
26. Feldman, G.M.; Mogyorosi, A.; Heck, G.L.; DeSimone, J.A.; Santos, C.R.; Clary, R.A.; Lyall, V. Salt-evoked lingual surface potential in humans. *J. Neurophysiol.* **2003**, *90*, 2060–2064. [CrossRef] [PubMed]
27. Katsumata, T.; Nakakuki, H.; Tokunaga, C.; Fujii, N.; Egi, M.; Phan, T.H.; Mummalaneni, S.; DeSimone, J.A.; Lyall, V. Effect of Maillard reacted peptides on human salt taste and the amiloride-insensitive salt taste receptor (TRPV1t). *Chem. Senses* **2008**, *33*, 665–680. [CrossRef] [PubMed]
28. Ogasawara, M.; Katsumata, T.; Egi, M. Taste properties of Maillard-reaction products prepared from 1000 to 5000 Da peptide. *Food Chem.* **2006**, *99*, 600–604. [CrossRef]
29. Coleman, J.; Williams, A.; Phan, T.H.; Mummalaneni, S.; Melone, P.; Ren, Z.; Zhou, H.; Mahavadi, S.; Murthy, K.S.; Katsumata, T.; et al. Strain differences in the neural, behavioral, and molecular correlates of sweet and salty taste in naive, ethanol- and sucrose-exposed P and NP rats. *J. Neurophysiol.* **2011**, *106*, 2606–2621. [CrossRef]
30. Fujimaki, M.; Arai, S.; Yamashita, M.; Kato, H.; Noguchi, M. Taste Peptide Fractionation from a Fish Protein Hydrolysate. *Agric. Biol. Chem.* **1973**, *37*, 2891–2898. [CrossRef]
31. Rhyu, M.R.; Song, A.Y.; Kim, H.Y.; Kim, S.S.; Tokunaga, C.; Phan, T.-H.T.; Heck, G.L.; DeSimone, J.A.; Lyall, V. Naturally occurring peptides in mature Korean Soy Sauce modulate TRPV1 Variant Salt Taste Receptor. In Proceedings of the Association for Chemoreception Sciences, St. Petersburg, FL, USA, 13–17 April 2011.
32. DeSimone, J.A.; Lyall, V.; Heck, G.L.; Phan, T.H.; Alam, R.I.; Feldman, G.M.; Buch, R.M. A novel pharmacological probe links the amiloride-insensitive NaCl, KCl, and NH(4)Cl chorda tympani taste responses. *J. Neurophysiol.* **2001**, *86*, 2638–2641. [CrossRef]
33. Lyall, V.; Heck, G.L.; Vinnikova, A.K.; Ghosh, S.; Phan, T.H.; Alam, R.I.; Russell, O.F.; Malik, S.A.; Bigbee, J.W.; DeSimone, J.A. The mammalian amiloride-insensitive non-specific salt taste receptor is a vanilloid receptor-1 variant. *J. Physiol.* **2004**, *558*, 147–159. [CrossRef]
34. Lyall, V.; Heck, G.L.; Vinnikova, A.K.; Ghosh, S.; Phan, T.H.; Desimone, J.A. A novel vanilloid receptor-1 (VR-1) variant mammalian salt taste receptor. *Chem. Senses* **2005**, *30* (Suppl. 1), i42–i43. [CrossRef] [PubMed]
35. Lyall, V.; Heck, G.L.; Phan, T.H.; Mummalaneni, S.; Malik, S.A.; Vinnikova, A.K.; Desimone, J.A. Ethanol modulates the VR-1 variant amiloride-insensitive salt taste receptor. II. Effect on chorda tympani salt responses. *J. Gen. Physiol.* **2005**, *125*, 587–600. [CrossRef]
36. Lyall, V.; Phan, T.H.; Mummalaneni, S.; Mansouri, M.; Heck, G.L.; Kobal, G.; DeSimone, J.A. Effect of nicotine on chorda tympani responses to salty and sour stimuli. *J. Neurophysiol.* **2007**, *98*, 1662–1674. [CrossRef]
37. Yamaguchi, S. The Synergistic Taste Effect of Monosodium Glutamate and Disodium 5′-Inosinate. *J. Food Sci.* **2006**, *32*, 473–478. [CrossRef]
38. Gunthorpe, M.J.; Rami, H.K.; Jerman, J.C.; Smart, D.; Gill, C.H.; Soffin, E.M.; Luis Hannan, S.; Lappin, S.C.; Egerton, J.; Smith, G.D.; et al. Identification and characterisation of SB-366791, a potent and selective vanilloid receptor (VR1/TRPV1) antagonist. *Neuropharmacology* **2004**, *46*, 133–149. [CrossRef]
39. Bachmanov, A.A.; Tordoff, M.G.; Beauchamp, G.K. Voluntary sodium chloride consumption by mice: Differences among five inbred strains. *Behav. Genet.* **1998**, *28*, 117–124. [CrossRef] [PubMed]
40. Meilgaard, M.; Civille, G.V.; Carr, B.T. Descriptive analysis Techniques. In *Sensory Evaluation Techniques*, 2nd ed.; Meilgaard, M., Civille, G.V., Carr, B.T., Eds.; CRC Press: Boca Raton, FL, USA, 1988; pp. 119–142.
41. Lyall, V.; Phan, T.H.; Mummalaneni, S.; Melone, P.; Mahavadi, S.; Murthy, K.S.; DeSimone, J.A. Regulation of the benzamil-insensitive salt taste receptor by intracellular Ca2+, protein kinase C, and calcineurin. *J. Neurophysiol.* **2009**, *102*, 1591–1605. [CrossRef] [PubMed]
42. Lyall, V.; Phan, T.H.; Ren, Z.; Mummalaneni, S.; Melone, P.; Mahavadi, S.; Murthy, K.S.; DeSimone, J.A. Regulation of the putative TRPV1t salt taste receptor by phosphatidylinositol 4,5-bisphosphate. *J. Neurophysiol.* **2010**, *103*, 1337–1349. [CrossRef]
43. Gu, X.F.; Lee, J.H.; Yoo, S.B.; Moon, Y.W.; Jahng, J.W. Intra-oral pre-treatment with capsaicin increases consumption of sweet solutions in rats. *Nutr. Neurosci.* **2009**, *12*, 149–154. [CrossRef]
44. Moon, Y.W.; Lee, J.H.; Yoo, S.B.; Jahng, J.W. Capsaicin receptors are colocalized with sweet/bitter receptors in the taste sensing cells of circumvallate papillae. *Genes Nutr.* **2010**, *5*, 251–255. [CrossRef]

45. Simon, S.A.; Gutierrez, R. TRP channels at the periphery of the taste and trigeminal systems, Chapter 7. In *Neurobiology of TRP Channels*, 2nd ed.; CRC Press: Boca Raton, FL, USA; Taylor & Francis: Abingdon, UK, 2017.
46. Kawashima, M.; Imura, K.; Sato, I. Topographical organization of TRPV1-immunoreactive epithelium and CGRP-immunoreactive nerve terminals in rodent tongue. *Eur. J. Histochem.* **2012**, *56*, e21. [CrossRef] [PubMed]
47. Huang, A.Y.; Wu, S.Y. Calcitonin Gene-Related Peptide Reduces Taste-Evoked ATP Secretion from Mouse Taste Buds. *J. Neurosci.* **2015**, *35*, 12714–12724. [CrossRef] [PubMed]
48. Dewis, M.L.; Phan, T.H.; Ren, Z.; Meng, X.; Cui, M.; Mummalaneni, S.; Rhyu, M.R.; DeSimone, J.A.; Lyall, V. N-geranyl cyclopropyl-carboximide modulates salty and umami taste in humans and animal models. *J. Neurophysiol.* **2013**, *109*, 1078–1090. [CrossRef] [PubMed]
49. Kim, M.J.; Son, H.J.; Kim, Y.; Kweon, H.J.; Suh, B.C.; Lyall, V.; Rhyu, M.R. Selective activation of hTRPV1 by N-geranyl cyclopropylcarboxamide, an amiloride-insensitive salt taste enhancer. *PLoS ONE* **2014**, *9*, e89062. [CrossRef]
50. Rhyu, M.R.; Lee, B.H.; Kim, Y.H.; Song, A.Y.; Son, H.J.; Oh, S.B.; Lyall, V.; Kim, E.Y. Soy-derived glycopeptides induce inward current in TRPV1-expressing cells by whole-cell patch-clamp recording. In Proceedings of the Association for Chemoreception Sciences, Sarasota, FL, USA, 25–29 April 2007.
51. Rhyu, M.-R.; Song, A.-Y.; Kim, E.-Y.; Oh, S.B.; Lee, Y.; Lim, W.C. Pharmaceutical Composition for Preventing/Treating TRPV1 Activity-Related and Inflammation-Related Diseases or Conditions Containing Maillard peptide Separated from well-Aged Traditional Soy Sauce as Active Ingredient. U.S. Patent No. 8,653,032 B2, 18 February 2014.
52. Treesukosol, Y.; Lyall, V.; Heck, G.L.; DeSimone, J.A.; Spector, A.C. A psychophysical and electrophysiological analysis of salt taste in Trpv1 null mice. *Am. J. Physiol. Regul. Integr. Comp. Physiol.* **2007**, *292*, R1799–R1809. [CrossRef]
53. Xu, J.J.; Elkaddi, N.; Garcia-Blanco, A.; Spielman, A.I.; Bachmanov, A.A.; Chung, H.Y.; Ozdener, M.H. Arginyl dipeptides increase the frequency of NaCl-elicited responses via epithelial sodium channel alpha and delta subunits in cultured human fungiform taste papillae cells. *Sci. Rep.* **2017**, *7*, 7483. [CrossRef]
54. Shigemura, N.; Shirosaki, S.; Ohkuri, T.; Sanematsu, K.; Islam, A.A.; Ogiwara, Y.; Kawai, M.; Yoshida, R.; Ninomiya, Y. Variation in umami perception and in candidate genes for the umami receptor in mice and humans. *Am. J. Clin. Nutr.* **2009**, *90*, 764s–769s. [CrossRef]
55. Shigemura, N.; Shirosaki, S.; Sanematsu, K.; Yoshida, R.; Ninomiya, Y. Genetic and molecular basis of individual differences in human umami taste perception. *PLoS ONE* **2009**, *4*, e6717. [CrossRef]

nutrients

MDPI

Article

Salt Taste Genotype, Dietary Habits and Biomarkers of Health: No Associations in an Elderly Cohort

Celeste Ferraris [1], Alexandria Turner [1], Kiranjit Kaur [1], Jessica Piper [1], Martin Veysey [2,3], Mark Lucock [1] and Emma L. Beckett [1,4,5,*]

[1] School of Environmental and Life Sciences, The University of Newcastle, Ourimbah, NSW 2258, Australia; celeste.ferraris@uon.edu.au (C.F.); Alexandria.Turner@uon.edu.au (A.T.); KiranjitKaur1@uon.edu.au (K.K.); Jessica.piper@uon.edu.au (J.P.); mark.lucock@newcastle.edu.au (M.L.)

[2] School of Medicine and Public Health, The University of Newcastle, Gosford, NSW 2250, Australia; martin.veysey@hyms.ac.uk

[3] Hull York Medical School, University of Hull, Heslington HU6 7RX, UK

[4] Hunter Medical Research Institute, New Lambton Heights, NSW 2305, Australia

[5] Priority Research Centre for Physical Activity and Nutrition, The University of Newcastle, Callaghan, NSW 2308, Australia

* Correspondence: emma.beckett@newcastle.edu.au

Received: 29 February 2020; Accepted: 8 April 2020; Published: 10 April 2020

check for updates

Abstract: A small amount of emerging research has observed variations between individual sensitivity, preference and intake of salt in the presence of single nucleotide polymorphisms (SNP) on the genes encoding salt taste receptors. Sodium intake is a significant risk factor for common diseases in elderly populations such as hypertension and cardiovascular disease; however, this does not fully explain the risk. Research into the influence of salt taste genetics on diet quality is yet to be undertaken and current research on indicators of health is limited and mixed in the direction of associations. Therefore, a secondary analysis of data from a well-characterised elderly cohort (the cross-sectional Retirement Health and Lifestyle Study, n = 536) was conducted to explore relationships between the salt taste-related SNP *TRPV1*-rs8065080 (assessed by Taqman genotyping assay), dietary habits and biomarkers of health. Data were analysed with standard least squares regression modelling and Tukey's HSD post hoc tests. No association was found between the *TRPV1*-rs8065080 genotype, sodium intake or multiple diet quality indices (assessed by food frequency questionnaire). Sodium-related markers of health including blood pressure and markers of kidney function (urinary creatinine and albumin/creatinine ratio) and general health markers, such as Body Mass Index (BMI), were also not related to *TRPV1*-rs8065080 genotype. To date, this study is the most comprehensive investigation conducted to determine if the *TRPV1*-rs8065080 genotype relates to sodium intake and health markers influenced by sodium intake. Although no significant relationships were found, these findings are an important contribution to the limited body of knowledge surround this SNP. In addition to further research across other ages and cultures, the *TRPV1*-rs8065080 genotype may interact with other ion channels, and so further studies are required to determine if polymorphic variations influence sodium intake, diet and health.

Keywords: salt; taste; *TRPV1* gene; rs806500; dietary; biomarker; elderly; nutrigenetics

1. Introduction

With an ageing population [1] and diet known to be a major modifiable determinant of disease risk, understanding the relationships between taste status, diet and health status may be important in detecting and managing at-risk groups [2]. Excess sodium intake in the elderly increases the risk of cardiovascular disease, hypertension [3], osteoporosis [4] and gastric cancers [5]. However, in most countries, salt intake levels remain higher than the World Health Organizations' recommendations [6]. Furthermore, sensitivity, preference and intake of salt vary widely between individuals. The influence of genetics on these factors has been demonstrated across each of the five tastes, including salt in limited early research [7,8]. Individual differences in salt taste perception have been attributed to variations on the genes encoding taste receptors [9,10]. Evidence for the role of genetics in salt taste and health is emerging, and therefore, many relationships remain to be characterised.

Salt taste is detected through ion channels [11]. Along with the epithelial sodium channel (ENaC), the transient receptor potential cation subfamily V member 1 (TRPV1) channel has been identified as a salt taste receptor that responds to a variety of cations [12]. The TRPV1 gene is located on chromosome 17 [13] and is expressed throughout the body [14]. Salt concentration levels elicit different taste pathways. Type 1 taste cells have been identified as activated when concentrations are appetitive [15], while Type II and III taste cells are triggered by higher salt concentrations that are perceived as aversive [15]. Furthermore, single nucleotide polymorphisms (SNP's) on the TRPV1 gene have been identified as having influence on the threshold levels at which individuals perceive salt solutions [9].

TRPV1-rs8065080 is a missense mutation with the single amino acid change from isoleucine to valine occurring at position 585 [13]. The frequency of alleles differs between populations. The *TRPV1*-rs8065080 T allele is more common in Caucasian, African and Hispanic populations, and the C allele more common in Asian populations [13]. Animal and cell culture studies demonstrate potential mechanistic roles for the *TRPV1*-rs8065080 polymorphism in ion channel function. Transfection of HeLa cells with the *TRPV1*-rs8065080 C allele results in a 20%–30% loss of channel function [16]. Furthermore, salt-sensitive rats on a high salt diet had reduced expression and function, with authors hypothesizing that this was a potential mechanism for salt sensitivity [17].

Human research on the function and dietary outcomes of the different *TRPV1*-rs8065080 variants is limited and mixed. In a study of ten TRPV1 SNPs (n = 95, white, young adults), *TRPV1*-rs8065080 was the only SNP identified as being related to salt suprathreshold taste sensitivity [9]; however, liking and salt intake were not assessed. In a sub-study of participants from the Guelph Family Study (n = 125) various TRPV1 SNPs, but not *TRPV1*-rs8065080, were found to be associated with higher sensitivity to, and preference for, salt [18]. Sodium intake was assessed by three TRPV1 SNP's (rs4790151, rs4790522 and rs877610); however, no significant relationships were found [18]. A smaller American study (n = 20), in which the *TRPV1*-rs8065080 T allele was associated with higher sensitivity to salt, found T allele carriers consumed higher amounts of sodium than C allele carriers [10]. However, the small cohort size in the study limits the power of the analysis. In addition to salt taste, a large Korean epidemiology study (n = 8842) found *TRPV1*-rs8065080 T allele carriers to have a higher preference for, and consumption of oily foods [19]. The varied relationships demonstrate SNP-related functional impacts on taste that require further definition.

The health outcomes of *TRPV1*-rs8065080 variants have been investigated in both sodium-related and non-sodium-related diseases. While sodium intake is an established risk factor in high blood pressure [3], a recent Taiwanese study did not find a connection to the TRPV1 SNP [20]. In male and female adults (n = 551), *TRPV1*-rs8065080 was not associated with systolic or diastolic blood pressure levels [20]. Conversely, the *TRPV1*-rs8065080 allele carriage has been associated with risk for other diseases not directly related to sodium intake, including type 2 diabetes and insulin sensitivity [19], the risk for knee osteoarthritis [21], cough and wheeze in asthmatics [16,22], and differential responses to pain [23,24].

The relationships between the *TRPV1*-rs8065080 polymorphism, salt intake and markers of health related to sodium intake remain to be fully elucidated. Therefore, we assessed the relationship between the *TRPV1*-rs8065080 genotype, sodium intake, diet quality, Body Mass Index (BMI), blood pressure and markers of kidney function in a well-characterised elderly cohort.

2. Materials and Methods

2.1. Subjects

This secondary cross-sectional analysis examined data from the Retirement Health and Lifestyle Study (RHLS) conducted on the NSW Central Coast of Australia from 2010 to 2012 [25]. Individuals living in private dwellings were randomly selected from extracts of the Wyong and Gosford local government areas Australian Commonwealth Electoral Rolls. Individuals from 12 participating retirement villages located in the same electorates were also randomly selected from retirement village resident lists. Participants were eligible to participate if they were aged 65 years or older and their primary residence for the last 12 months or more was located within the Wyong or Gosford local government areas. Those who were not living independently or were residing in a communal setting other than a retirement village, had been living in the area for less than 12 months, and/or were in the process of relocating, were not eligible to participate. Individuals were also ineligible if another member of their household was taking part in the study. People with language or other communication difficulties, who were cognitively impaired or unable to provide informed consent, were also excluded [26]. Participants were not excluded based on pre-existing health conditions. In total, 831 people were recruited for this study, however, the provision of a blood sample for genotyping was optional. Only those who were successfully genotyped for *TRPV1*-rs8065080 and provided a valid food frequency questionnaire were included in this sub-study. Complete data sets were available for 536 participants. Ethics approval for the RHLS study was granted by The Human Research Ethics Committee of the University of Newcastle (Reference No. H-2008-0431) and written consent obtained from participants [27].

2.2. Anthropometric Measures

Age, sex, education and income levels were collected via questionnaires. Anthropometric measures followed the International Society for the Advancement of Kinanthropometry (ISAK) guidelines [28]. Participants wore light comfortable clothing and measures were repeated until two consecutive values within 0.5 cm were recorded. The stretch stature method was used to measure height and recorded to the nearest 0.01 cm [28]. Digital scales (Wedderburn© UWPM150 Platform Scale) measured weight which was recorded to the nearest 0.01 kg. Calculations using the height and weight measures determined Body Mass Index for each participant (BMI = weight (kg)/height (m^2)). Waist, hip and waist to hip ratio measures were also taken, following the ISAK guidelines [28].

2.3. Blood Pressure Readings

Blood pressure (BP) measurements were administered by qualified clinical staff using an OMRON IA2 machine. Readings were taken from both arms, allowing at least one minute between measurements. In the arm with the highest reading, a further two BP measurements were taken and recorded, these two consecutive measurements were averaged and used in statistical analyses. If the consecutive readings differed by >10 mmHg for systolic blood pressure or by >6 mmHg for diastolic blood pressure, a further fourth measurement was taken. Measurement was abandoned after a maximum of four readings (on one arm). Participants were excluded if their measurement was invalid (consecutive BP readings differed by more than 10 mmHg systolic or 6 mmHg diastolic); there was a physical limitation preventing measurement; they presented with a very high BP curtailing measurement, or there was a machine error.

2.4. Collection of Biological Samples

After fasting, whole blood was collected via venipuncture, by a trained nurse, into EDTA-lined tubes and stored at −20 °C [27]. Urine samples were also collected without preservative on the morning of the clinic visit while the participant was fasting. Urinalysis was conducted by the Hunter Area Pathology Service.

2.5. Genotyping

In the RHLS study, deoxyribonucleic acid (DNA) was isolated from peripheral blood cells with the QIAGEN QIAmp DNA mini-kit following the manufacturers' protocol [25,29]. The DNA samples were stored at −20 °C [25]. The SNP was assessed with quantitative polymerase chain reaction (qPCR) in the QuantStudio 7 Flex Real-Time PCR System [30]. Allelic discrimination for *TRPV1*-rs8065080 was performed using TaqMan™ assay code C___11679656_10 (Applied Biosystems™, ThermoFisher Scientific, CA, USA). In 384 well plates (MicroAmp® Optical 384-Well Reaction Plate with Barcode, Applied Biosystems™, ThermoFisher Scientific, CA, USA), 2.25 µL of DNA was dried down in each well. Each qPCR reaction contained the DNA and 2.50 µL 2x TaqMan™ Master Mix, 0.25 µL 20x TaqMan™ Assay and 2.25 µL nuclease-free water (UltraPure™ Distilled Water, Invitrogen, ThermoFisher Scientific, CA, USA). Two no-template controls with all components except DNA were run for specificity validation. The plates were sealed (MicroAmp® Optical Adhesive Film, Applied Biosystems™, ThermoFisher Scientific, CA, USA) and centrifuged (SelectSpin™ Plate Centrifuge, Select BioProducts, NJ, USA). Denaturing occurred for 10 min at 95 °C initially, then 15 s at 95 °C over 40 cycles. Elongation and annealing ran for 10 min at 60 °C. Data were captured at the end of each cycle.

2.6. Dietary Assessment

A food frequency questionnaire (FFQ), containing 225 items, and covering all food groups [31] was used to estimate daily sodium intake and to calculate the diet quality indices. Data were analysed using Foodworks 2.10.146 (Xyris Software, Brisbane, QLD, Australia) [32]. Three diet quality indices were calculated from the available data: The Dietary Guideline Index (DGI) [33], the Australian Recommended Food Score (ARFS) [34,35] and the Australian Healthy Eating Index (Aust-HEI) [36].

The DGI is a 150-point index based on the Dietary Guidelines for Australian Adults, Australian Guide to Healthy Eating, national indicators for food and nutrition and the Australian Alcohol Guidelines [33,37,38]. A score of 0–10, proportionally adjusted, was allocated across 15 food categories, with higher scores indicating higher quality dietary intake [33]. The fifteen constituents of the DGI are set to assess a participant's intake of key nutrients from core food groups, the proportion of key nutrient intakes from healthy food types (e.g., lean meats or wholegrain cereals), diversity of foods in the diet and intakes of unhealthy foods.

The ARFS reflects the Australian Dietary Guidelines and focuses on dietary variety as an indicator of diet quality [34,35]. Points are allocated within eight food sub-scales (vegetables, fruits, protein sources, grains, dairy, fats and alcohol). Points are assigned based on of frequency of consumption reflective of the guidelines. The ARFS score was calculated by summing the points for each item. A maximum of 74 points were possible, with higher scores indicating a wide variety in dietary intakes [35]. Minor modifications were made to the ARFS scorings system due to differences in the foods listed on the FFQ used. Garlic, beetroot and zucchini were not listed on the FFQ used here, but cucumber, asparagus and sweet potato were and so were substituted in the vegetable sub-scale. Similarly, frozen or canned fruits and pears were substituted for berries and kiwifruit/plums/grapes in the fruit sub-scale. In the grains sub-scale, the all-bran was replaced with all high fibre cereals not already captured. In the protein foods sub-scale, veal was not listed in the FFQ, but turkey/quail/duck were and so they were substituted. Alcohol consumption was recorded as an average so consumption per sitting could not be calculated. Points were awarded based on average annual consumption equivalent to 1 or 2 drinks per day for a maximum of four days of the week. Zero points were awarded to those who consumed more than this or never consumed alcohol.

The Aust-HEI generates a score for healthy dietary behaviours and food consumption, equally weighting dietary variety, measures of healthy choices, fruit and vegetable consumption, fat consumption and consumption of discretionary foods [36]. Possible scores range from 0–60 points.

2.7. Medical History and Medication Status

Self-reported medical history of cardiovascular diseases (hypertension, heart disease, stroke, heart attack and vascular disease) and kidney disease were recorded via interviewer administered questionnaire. Regular prescription medication use was recorded via presentation of medication packages to investigators of photographic recording of brand, active ingredient and dose, and self-report of frequency of use via interviewer administered surveys.

2.8. Statistical Analysis

Data analyses were completed with statistical analysis software JMP (Version 14.2; SAS Institute Inc., USA). Participant characteristics were reported as number and percentage of the total cohort when variables were categorical and mean, minimum and maximum ranges and standard deviation (SD) when continuous. Genotype allele frequency was reported as number and percentage with distributions analyzed using Pearson's chi-squared tests. Associations between genotype and the continuous variables were assessed with standard least squares regression. Least squared means were compared with appropriate adjustments, or raw means unadjusted, using Tukey's HSD post hoc test. The threshold for statistical significance was $p = <0.05$. Stratified analyses by each of the potential confounding variables are also presented where appropriate.

3. Results

3.1. Participant Characteristics

After exclusions, data for 536 participants were available for analysis. Participant age ranged from 65 to 94 years (mean 77.4 years, standard deviation of 6.8 years, Table 1). The mean BMI of participants was in the overweight range (28.5; Table 1). The mean DGI was 97/150, the mean ARFS was 26.8/74 and the mean AUST-HEI was 30.2/60. Mean estimated sodium intake was 2052 mg/day.

Table 1. Distribution of continuous variables.

Variable	Mean	Minimum	Maximum	SD
Age (years)	77.4	65	94	6.8
DGI (150-point index)	97.0	30.9	132.6	15.8
ARFS (74-point index)	28.9	6	50	7.6
AUST-HEI (60-point index)	30.2	5	50	9.7
BMI (kg/m^2)	28.5	17.1	46.3	4.8
Sodium intake (mg/day)	2052	506	8250	843

BMI: Body Mass Index.

All dietary indices and BMI were normally distributed. 45% of participants were male. The majority reported incomes in the middle-income bracket with education levels of TAFE qualification or higher (Table 2). Due to the small numbers of participants in the higher income bracket, this variable was collapsed into two groups for analysis (above AUD 20,000 and below AUD 20,000). Similarly, ex-smokers and current smokers were collapsed into "ever" smokers for analysis.

Table 2. Distribution of categorical variables.

Variable	n	%
Sex		
Males	241	45.0
Females	295	55.0
Income		
<AUD 20,000 per year	165	31.5
AUD 20,000 to AUD 60,000 per year	322	61.6
>AUD 60,000 per year	36	6.9

Table 2. *Cont.*

Variable	n	%
Education		
≤Trade qualification	177	33.1
TAFE or other certificates	295	55.1
≥Bachelor degree	63	11.8
Smoking		
Current smoker	15	2.8
Ex-smoker	253	47.2
Never smoked	268	50.0
History of cardiovascular disease [a,b]		
Yes	219	40.9
No	317	59.1
History of kidney disease [b]		
Yes	90	17.5
No	425	82.5
Use of anti-hypertensive medication [b]		
Yes	118	22.0
No	418	78.0
Regular use of any prescription medication [b]		
Yes	438	81.7
No	98	18.3

[a] includes hypertension, heart disease, stroke, heart attack and vascular disease; [b] self-reported.

3.2. Genotype Distributions

The variant *TRPV1*-rs8065080 allele (C) had a frequency of 0.36, and the ancestral allele (T) had a frequency of 0.64. The heterozygous genotype (C/T) was the most common, followed by T/T and C/C (Table 3). The distributions of age (p = 0.5), sex (p = 0.7), income (p = 0.5), education (p = 0.6), history of cardiovascular disease (p = 0.2), history of kidney disease (p = 0.5), use of antihypertensive medication (p = 0.1) and use of any prescription medication (p = 0.1) did not vary significantly by *TRPV1*-rs8065080 genotype.

Table 3. *TRPV1*-rs8065080 genotype distribution.

SNP	Female *n* (%)	Male *n* (%)	Total *n* (%)
C/C	33 (11.2)	33 (13.7)	66 (12.3)
C/T	134 (45.4)	116 (48.1)	250 (46.6)
T/T	128 (43.4)	92 (38.2)	220 (41.1)

3.3. Sodium Intake by Genotype

As TRPV1 is involved in detecting salt taste, the relationship between sodium intake and *TRPV1*-rs8065080 genotype was assessed. Sodium intake was higher in males than females with mean intakes of 2226 ± SD814 mg/day and 1910 ± SD840 ($p < 0.001$) and was higher in the higher income group (1906 ± SD924 mg/day vs. 2129 ± SD8044 mg/day, p = 0.005). Intake also reduced with age ($\beta = -0.13$, p = 0.004). However, sodium intake was not related to smoking status (p = 0.9), education (p = 0.9), history of cardiovascular disease (p = 0.9, history of kidney disease (p = 0.3), use of anti-hypertensive medication (p = 0.5) or use of any prescription medication (p = 0.6). Therefore, sodium intake was assessed by genotype in the complete cohort without adjustments, and with adjustments for age, income and sex. However, there were no significant differences in sodium intake by genotype (Figure 1). The association remained non-significant when salt intake was categorically analysed by quartiles (χ^2 =4.8, p = 0.3). Furthermore, there were no significant differences in salt intake by genotype when analyses were stratified by each of the potential adjustment variables (Table S1).

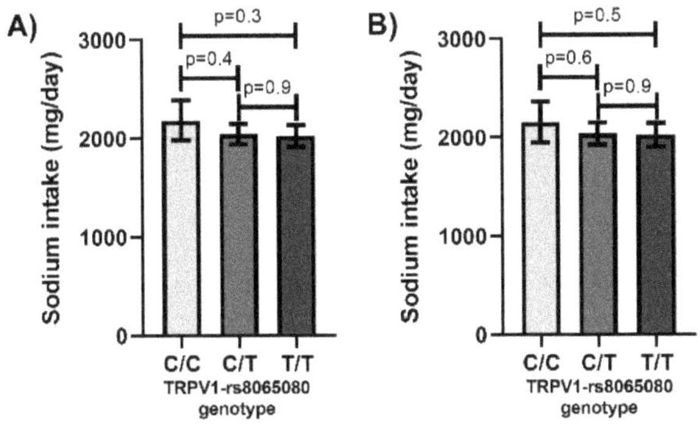

Figure 1. Sodium intake does not vary by TRPV1-rs8065080 genotype. (**A**) Unadjusted mean values (**B**) Least- squares means with adjustments for age, income and sex. Error bars mark 95% confidence intervals.

3.4. *Diet Quality by Genotype*

It is often hypothesized that taste genetics are involved in modulating dietary preferences and intake. Therefore, we used three diet quality indices to assess the relationship between *TRPV1*-rs8065080 genotype and diet quality. DGI and AUST-HEI were higher in females than males and both increased with education (Table S2). The DGI was also higher in those who had never smoked, compared to those with a history of smoking (Table S2). ARFS did not vary by sex, education or smoking status (Table S2). None of the diet quality indices varied by income (Table S2), age (DGI p = 0.1, AUST-HEI p = 0.2, ARFS p = 0.2), history of cardiovascular disease (DGI p = 0.2, AUST-HEI p = 0.2, ARFS p = 0.3), history of kidney disease (DGI p = 0.4, AUST-HEI p = 0.2, ARFS p = 0.2), use of anti-hypertensive medication (DGI p = 0.3, AUST-HEI p = 0.3, ARFS p = 0.1) or use of any prescription medication (DGI p = 0.2, AUST-HEI p = 0.1, ARFS p = 0.1). Therefore, the relationship between diet quality and *TRPV1*-rs8065080 genotype was assessed without adjustments, and with adjustments for sex, education and smoking. However, there were no significant differences in any of the diet quality indices by genotype (Figure 2) Furthermore, there were no significant differences in any of the indices by genotype when analyses were stratified by each of the potential adjustment variables (Table S1).

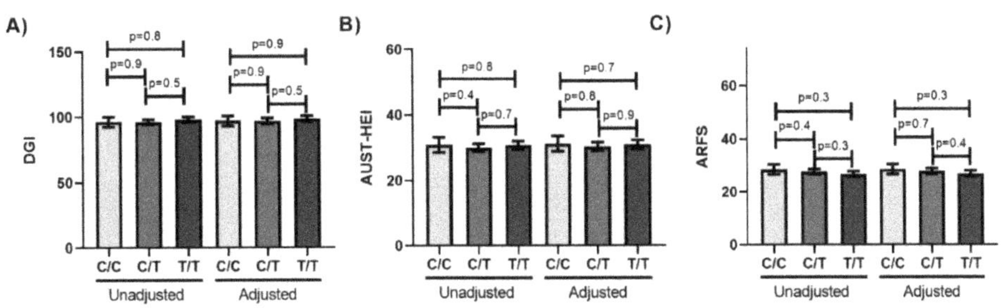

Figure 2. Diet quality indices do not vary by TRPV1-rs8065080 genotype. (**A**) DGI (**B**) AUST-HEI (**C**) ARFS. Error bars mark 95% confidence intervals. Unadjusted present mean values with no adjustments. Adjusted present Least-squares means with adjustments for sex, education and smoking.

3.5. *Markers of Health by Genotype*

The relationship between *TRPV1*-rs8065080 genotype and BMI was assessed as a marker of health status. BMI reduced with age (β = −0.17, p = 0.0001), was higher in those who had a history of smoking (29.1 ± 5.1 vs 27.7 ± 5.6, p = 0.005), and reduced with increasing education (β = 0.12, p = 0.03) and income (β =0.13, 0.01). BMI did not vary by sex (p = 0.7) history of cardiovascular disease (p = 0.1), history of kidney disease (p = 0.4), use of anti-hypertensive medication (p = 0.2) or use of any prescription medication (p = 0.1). Therefore, the relationship between BMI and *TRPV1*-rs8065080 genotype was assessed without adjustments, and with adjustments for age, smoking status, education and income. However, BMI did not vary by genotype (Figure 3A).

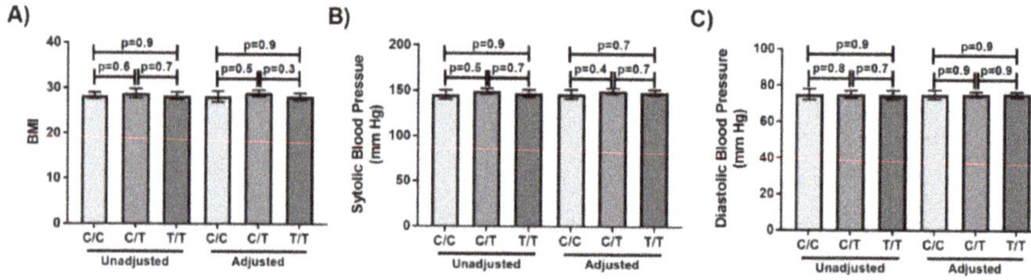

Figure 3. BMI, systolic blood pressure and diastolic blood pressure do not vary by TRPV1-rs8065080 genotype in the elderly. (**A**) BMI (**B**) Systolic blood pressure (**C**) Diastolic blood pressure. Error bars mark 95% confidence intervals. Unadjusted present mean values with no adjustments. Adjusted present Least-squares means with adjustments for age, smoking status, education and income.

The relationship between *TRPV1*-rs8065080 genotype and blood pressure was assessed as a marker of a salt-sensitive health outcome. Systolic and diastolic blood pressure both increased with age (p = 0.003 and p = 0.0001, respectively), and were higher in males (p = 0.003 and p = 0.006, respectively) and those who had a history of smoking (p = 0.03 and p = 0.02, respectively). Blood pressure did not vary by education (p = 0.9), income (p = 0.9) history of cardiovascular disease (p = 0.2), history of kidney disease (p = 0.1), use of anti-hypertensive medication (p = 0.2) or use of any prescription medication (p = 0.1). Therefore, the relationship between blood pressure and the *TRPV1*-rs8065080 genotype was assessed without adjustments, and with adjustments for age, smoking status, education and income. However, neither systolic nor diastolic blood pressure varied by genotype (Figure 3B,C). The incidence of hypertension did not vary between genotypes (χ = 0.5, p = 0.8). Results did not vary when adjustment for sodium intake was added, and there was no significant interaction between *TRPV1*-rs8065080 genotype and sodium intake in predicting systolic or diastolic blood pressure ($p_{interaction}$ = 0.6 and 0.7, respectively). Furthermore, there were no significant differences these markers by genotype when analyses were stratified by each of the potential adjustment variables (Table S1).

The relationship between the *TRPV1*-rs8065080 genotype and urine creatinine and albumin to creatinine ratio was assessed as a marker of kidney health. Urine creatinine was inversely related to age (β = −0.14. p = 0.001), was higher in males (8.5 ± SD5.0, v 10.2 ± SD5.2, mmol/L, p = 0.0006) and those with a history of kidney disease (8.9 ± SD2.0, v 10.0 ± SD4.2, mmol/L, p = 0.02). Urine markers did not vary by education, income, smoking history, history of cardiovascular disease (p = 0.9), use of anti-hypertensive medication (p = 0.2) or use of any prescription medication (p = 0.3). Therefore, the relationship between urine markers and *TRPV1*-rs8065080 genotype was assessed without adjustments, and with adjustments for age and sex. However, neither urine creatinine nor the albumin to creatinine ratio varied by genotype (Table 4). Results did not vary when adjustment for sodium intake was added, and there was no significant interaction between *TRPV1*-rs8065080 genotype and sodium intake in urine creatinine, albumin or albumin to creatinine ratio ($p_{interaction}$ = 0.7 and 0.5, respectively). Furthermore, there were no significant differences in salt intake by genotype when analyses were stratified by each of the potential adjustment variables (Table S1).

Table 4. *TRPV1*-rs8065080 and urine creatinine and albumin to creatinine ratio.

	Unadjusted (Mean ± SEM)				Adjusted (Mean ± SEM)			
	C/C	C/T	T/T	p_{trend}	C/C	C/T	T/T	p_{trend}
Creat (mmol/L)	9.3 ± 0.6	9.7 ± 0.3	9.2 ± 0.3	0.5	9.1 ± 0.7	9.8 ± 0.4	9.2 ± 0.4	0.4
Alb/Creat Ratio (mg/mmol)	2.5 ± 4.0	9.3 ± 2.0	3.6 ± 2.3	0.1	5.7 ± 4.0	9.7 ± 2.3	7.5 ± 2.5	0.2

4. Discussion

This study is the most comprehensive characterisation of the relationships between the *TRPV1*-rs8065080 genotype, sodium intake and health markers influenced by sodium intake to date. Despite the suggested role for *TRPV1* in the detection of salts [12] including sodium [9,10], no association was found between the SNP, sodium intake or sodium-related markers of health including blood pressure and markers of kidney function in the elderly. Additionally, although salt preference potentially influences dietary patterns and quality, BMI and multiple diet quality indices were not related to the *TRPV1*-rs8065080 genotype.

The data presented here, in elderly participants, supports the finding for *TRPV1*-rs4790151, -rs4790522 and -rs877610 and sodium intake in a previous smaller study (n = 125) which found that these SNPs were not associated with sodium intake in adults and preschool-aged children. However, it is contradictory to the data presented in a small study (n = 20) of young predominantly Caucasian participants by Pilic et al. [10], who reported that the *TRPV1*-rs8065080 T allele carriers consumed higher amounts of sodium than C allele carriers. From an SNP perspective, the *TRPV1*-rs8065080 missense mutation alters one amino acid and, therefore, direct comparison is only possible between this study and the work of Pilic et al. [10]. The contradictory results may be explained by the significant difference in ages between the two cohorts, as age has been found to be a factor in phenotypical variance in genetic expression [39]. However, the limited research available in the genetics of salt taste restricts the valuable comparative analysis, highlighting the need for studies such as this that contribute to the characterisation of TRPV1 SNPs.

Despite no relationship between the presence of the SNP and sodium intake being found, assessing the relationship between *TRPV1*-rs8065080 genotype, blood pressure and kidney function remained important due to the extra-oral expression of TRPV1 throughout the body [14]. Extra-oral taste receptors act as chemosensors [40] and therefore may have biological impacts independent of the modulation of diet via taste thresholds and preferences. However, no association was found between blood pressure and the *TRPV1*-rs8065080 genotype in the present study. This supports the findings of a similarly sized study in male and female Taiwanese adults (n = 551) which found no association between *TRPV1*-rs8065080 and systolic or diastolic blood pressure levels [20]. Together, these two well-sized studies suggest that the *TRPV1*-rs8065080 genotype does not directly relate to blood pressure in adulthood nor in later life of the elderly. Demonstrating the presence or absence of this relationship in the elderly is important, as ageing is known to heighten the effect of sodium intake on blood pressure [41]. The present study is the first to assess the relationship between the *TRPV1*-rs8065080 genotype and markers of kidney function; therefore, the absence of association is novel.

The relationship between the *TRPV1*-rs8065080 genotype and broader markers of dietary composition, including BMI and diet quality indices, has not been explored previously. This approach was taken to reflect overall dietary composition, as dietary patterns are important in determining disease risk. Furthermore, a nutrient-focused reductionist approach may miss important and relevant associations. However, the *TRPV1*-rs8065080 genotype was not related to diet quality by any of the indices used and was not related to BMI.

Strengths of this study include the large-sized and well-characterised nature of the cohort that is representative of the larger population. The *TRPV1*-rs8065080 allele frequencies in participants matched those found in global populations [13]. The mean BMI (28.5 kg/m^2) placed the study population in the overweight category, and this was reflective of the national average for older adults [42]. The average sodium intake reported in this cohort was also similar to the broader Australian cohort [43]. Furthermore, multiple outcome variables were able to be assessed in a single cohort as a result of the study population being well-characterised.

Limitations also need to be considered in the interpretation of these data. As a secondary analysis of data from a cross-sectional study, time-sensitive outcomes such as the age-related decline in taste [44] cannot be accounted for. As the cohort was 65 years and older, the results are not necessarily generalizable to the wider adult or youth population. However, the elderly cohort is suitable for the research question, as genetic and dietary exposures, and their interactions, accumulate over a lifetime and the relationships may only become apparent in older age. Dietary data collected with FFQs may be subject to reporting bias, including inaccurate recall, an under-reporting of quantities of discretionary food and over-reporting of healthful foods. Added salt used as seasoning is also likely to be under-reported via FFQs. However, the use of multiple dietary indices rather than a focus on estimated nutrient intakes improves the integrity of the body of data presented here. Future studies using 24-h urinary sodium should be conducted to confirm this result, as this would more precisely demonstrate intake. While the relationship between the *TRPV1*-rs8065080 genotype and the outcome variables are well-investigated in the present study, and limited associations found, interactions with other *TRPV1* SNPs or other ion channel taste receptor SNPs may exist. For this reason, polymorphic variant studies are needed in the future.

5. Conclusions

Additional studies are needed in more diverse age and cultural groups to determine if the lack of associations found between the *TRPV1*-rs8065080 genotype with sodium intake and markers of health is confined to the Caucasian elderly demographic. Furthermore, while this study was the first to characterise the relationship between *TRPV1*-rs8065080 genotype and markers of dietary intake, quality and health outcomes—data on perception, preference and sensitivity were not available, and further studies are needed to elucidate the differential impacts of oral versus extra-oral taste receptors. However, the extensive characterisation presented here will be important in the interpretation of the results of future studies assessing the relationship between the *TRPV1*-rs8065080 genotype, and dietary and health outcomes.

Supplementary Materials: The following are available online at http://www.mdpi.com/2072-6643/12/4/1056/s1, Table S1: Stratified analyes, Table S2: Diet quality indices with sex, income, education and smoking.

Author Contributions: Conceptualization, C.F. and E.L.B.; Data curation, C.F., A.T., K.K., J.P., M.L. and E.L.B.; Formal analysis, C.F. and E.L.B.; Funding acquisition, M.V., M.L. and E.L.B.; Investigation, C.F., A.T., K.K., J.P., M.V., M.L. and E.L.B.; Methodology, C.F. and E.L.B.; Project administration, M.V., M.L. and E.L.B.; Resources, M.V., M.L. and E.L.B.; Writing—original draft, C.F. and E.L.B.; Writing—review and editing, C.F., A.T., K.K., J.P., M.V., M.L. and E.L.B. All authors have read and agreed to the published version of the manuscript

Funding: This research was conducted as part of the Retirement Health and Lifestyle Study, with initial and ongoing funding provided by the Australian Research Council (G0188386), Central Coast Local Health District Public Health Unit (G0190658/G1700259), UnitingCare Ageing NSW/ACT (G0189230), Urbis Pty Ltd. (G0189232), Valhalla Village Pty Ltd. (G1000936), and Hunter Valley Research Foundation.

Acknowledgments: Authors acknowledge the role of Paul Roach in the design of the original study and the roles of Charlotte Martin, Zoe Yates, Katrina King and Suzanne Niblett in sample collection and data management.

Conflicts of Interest: The authors declare no conflicts of interest.

References

1. Australian Bureau of Statistics. *Twenty Years of Population Change*; ABS: Canberra, Australia, 2019. Available online: https://www.abs.gov.au/ausstats/abs@.nsf/0/1CD2B1952AFC5E7ACA257298000F2E76 (accessed on 2 February 2020).

2. Australian Institute of Health and Welfare. *Nutrition Across the Life Stages*; AIHW: Canberra, Australia, 2018. Available online: https://www.aihw.gov.au/getmedia/fc5ad42e-08f5-4f9a-9ca4-723cacaa510d/aihw-phe-227.pdf.aspx?inline=true (accessed on 21 January 2020).
3. Strazzullo, P.; D'Elia, L.; Kandala, N.-B.; Cappuccio, F.P. Salt intake, stroke, and cardiovascular disease: Meta-analysis of prospective studies. *BMJ* **2009**, *339*, b4567. [CrossRef] [PubMed]
4. Frassetto, L.A.; Morris, R.C., Jr.; Sellmeyer, D.E.; Sebastian, A. Adverse effects of sodium chloride on bone in the aging human population resulting from habitual consumption of typical American diets. *J. Nutr.* **2008**, *138*, 419S–422S. [CrossRef] [PubMed]
5. Yang, W.G.; Chen, C.B.; Wang, Z.X.; Liu, Y.P.; Wen, X.Y.; Zhang, S.F.; Sun, T.W. A case-control study on the relationship between salt intake and salty taste and risk of gastric cancer. *World J. Gastroenterol.* **2011**, *17*, 2049–2053. [CrossRef] [PubMed]
6. Thout, S.R.; Santos, J.A.; McKenzie, B.; Trieu, K.; Johnson, C.; McLean, R.; Arcand, J.; Campbell, N.R.C.; Webster, J. The science of salt: Updating the evidence on global estimates of salt intake. *J. Clin. Hypertens.* **2019**, *21*, 710–721. [CrossRef] [PubMed]
7. Garcia-Bailo, B.; Toguri, C.; Eny, K.M.; El-Sohemy, A. Genetic variation in taste and its influence on food selection. *J. Int. Biol.* **2008**, *13*, 69–80. [CrossRef] [PubMed]
8. Chamoun, E.; Mutch, D.M.; Allen-Vercoe, E.; Buchholz, A.C.; Duncan, A.M.; Spriet, L.L.; Haines, J.; Ma, D.W.L. A review of the associations between single nucleotide polymorphisms in taste receptors, eating behaviors, and health. *Crit. Rev. Food Sci. Nutr.* **2018**, *58*, 194–207. [CrossRef]
9. Dias, A.G.; Rousseau, D.; Duizer, L.; Cockburn, M.; Chiu, W.; Nielsen, D.; El-Sohemy, A. Genetic variation in putative salt taste receptors and salt taste perception in humans. *Chem. Sens.* **2012**, *38*, 137–145. [CrossRef]
10. Pilic, L.; Mavrommatis, Y. Genetic predisposition to salt-sensitive normotension and its effects on salt taste perception and intake. *Br. J. Nutr.* **2018**, *120*, 721–731. [CrossRef]
11. Roper, S.D. The taste of table salt. *Euro. J. Phys.* **2015**, *467*, 457–463. [CrossRef]
12. Desimone, J.A.; Lyall, V. Taste receptors in the gastrointestinal tract III. Salty and sour taste: Sensing of sodium and protons by the tongue. *Am. J. Physiol. Gastro. Liver. Physiol.* **2006**, *291*, G1005–G1010. [CrossRef]
13. National Center for Biotechnology Information. *Database of Single Nucleotide Polymorphisms (dbSNP)*. Available online: https://www.ncbi.nlm.nih.gov/snp/ (accessed on 7 January 2019).
14. Uhlen, M.; Fagerberg, L.; Hallström, B.M.; Lindskog, C.; Oksvold, P.; Mardinoglu, A.; Sivertsson, Å.; Kampf, C.; Sjöstedt, E.; Asplund, A.; et al. Tissue-based map of the human proteome. *Science* **2015**, *347*, 1260419. [CrossRef] [PubMed]
15. Oka, Y.; Butnaru, M.; von Buchholtz, L.; Ryba, N.J.; Zuker, C.S. High salt recruits aversive taste pathways. *Nature* **2013**, *494*, 472–475. [CrossRef] [PubMed]
16. Cantero-Recasens, G.; Gonzalez, J.R.; Fandos, C.; Duran-Tauleria, E.; Smit, L.A.M.; Kauffmann, F.; Antó, J.M.; Valverde, M.A. Loss of function of transient receptor potential vanilloid 1 (TRPV1) genetic variant is associated with lower risk of active childhood asthma. *J. Biol. Chem.* **2010**, *285*, 27532–27535. [CrossRef] [PubMed]
17. Spradley, F.T.; Pollock, D.M.; Pollock, J.S. Dahl salt-sensitive rats on a high-fat diet develop hypertension and enhanced constriction to angiotensin II without changing endothelial-dependent vasorelaxation. *FASEB J.* **2010**, *24*, 1025–1029. [CrossRef]
18. Chamoun, E.; Carroll, N.; Duizer, L.; Qi, W.; Feng, Z.; Darlington, G.; Duncan, A.; Haines, J.; Ma, D. The relationship between single nucleotide polymorphisms in taste receptor genes, taste function and dietary intake in preschool-aged children and adults in the Guelph family health study. *Nutrients* **2018**, *10*, 990. [CrossRef] [PubMed]
19. Park, S.; Zhang, X.; Lee, N.R.; Jin, H.S. TRPV1 gene polymorphisms are associated with Type 2 Diabetes by their interaction with fat consumption in the Korean genome epidemiology study. *J. Nutrigenet. Nutrigenom.* **2016**, *9*, 47–61. [CrossRef]
20. Er, L.-K.; Teng, M.-S.; Wu, S.; Hsu, L.-A.; Tzeng, I.S.; Cheng, C.-F.; Chang, H.-I.; Chou, H.-H.; Ko, Y.-L. Combined effect of acid-sensing ion channel 3 and transient receptor potential vanilloid 1 gene polymorphisms on blood pressure variations in Taiwanese. *Ci Ji Yi Xue Za Zhi* **2018**, *30*, 29–36. [CrossRef]
21. Valdes, A.M.; De Wilde, G.; Doherty, S.A.; Lories, R.J.; Vaughn, F.L.; Laslett, L.L.; Maciewicz, R.A.; Soni, A.; Hart, D.J.; Zhang, W.; et al. The Ile585Val TRPV1 variant is involved in risk of painful knee osteoarthritis. *Ann. Rheum. Dis.* **2011**, *70*, 1556–1561. [CrossRef]

22. Smit, L.A.; Kogevinas, M.; Anto, J.M.; Bouzigon, E.; Gonzalez, J.R.; Le Moual, N.; Kromhout, H.; Carsin, A.E.; Pin, I.; Jarvis, D.; et al. Transient receptor potential genes, smoking, occupational exposures and cough in adults. *Respir. Res.* **2012**, *13*, 26. [CrossRef]
23. Binder, A.; May, D.; Baron, R.; Maier, C.; Tolle, T.R.; Treede, R.D.; Berthele, A.; Faltraco, F.; Flor, H.; Gierthmuhlen, J.; et al. Transient receptor potential channel polymorphisms are associated with the somatosensory function in neuropathic pain patients. *PLoS ONE* **2011**, *6*, e17387. [CrossRef]
24. Kim, H.; Neubert, J.K.; San Miguel, A.; Xu, K.; Krishnaraju, R.K.; Iadarola, M.J.; Goldman, D.; Dionne, R.A. Genetic influence on variability in human acute experimental pain sensitivity associated with gender, ethnicity and psychological temperament. *Pain* **2004**, *109*, 488–496. [CrossRef] [PubMed]
25. Beckett, E.L.; Duesing, K.; Martin, C.; Jones, P.; Furst, J.; King, K.; Niblett, S.; Yates, Z.; Veysey, M.; Lucock, M. Relationship between methylation status of vitamin D-related genes, vitamin D levels, and methyl-donor biochemistry. *J. Nutri. Intemed. Met.* **2016**, *6*, 8–15. [CrossRef]
26. Travers, C.; Dixon, A.; Laurence, A.; Niblett, S.; King, K.; Lewis, P.; Owen, N.; Veysey, M. Retirement Health and Lifestyle Study: Australian neighborhood environments and physical activity in older adults. *Environ. Beh.* **2018**, *50*, 426–453. [CrossRef]
27. Beckett, E.L.; Martin, C.; Boyd, L.; Porter, T.; King, K.; Niblett, S.; Yates, Z.; Veysey, M.; Lucock, M. Reduced plasma homocysteine levels in elderly Australians following mandatory folic acid fortification—A comparison of two cross-sectional cohorts. *J. Nutri. Intermed. Met.* **2017**, *8*, 14–20. [CrossRef]
28. Marfell-Jones, M.; Norton, K.; Carter, L.; Olds, T. *International Standards for Anthropometric Assessment*; International Society for the Advancement of Kinanthropometry: South Australia, 2001. Available online: https://docplayer.net/25335980-International-standards-for-anthropometric-assessment.html (accessed on 28 January 2020).
29. QIAGEN. *QIAamp®DNA Mini and Blood Mini Handbook*, 3rd ed. 2012. Available online: https://www.qiagen.com/au/resources/resourcedetail?id=62a200d6-faf4-469b-b50f-2b59cf738962&lang=en (accessed on 18 February 2020).
30. ThermoFisher Scientific. *QuantStudio 7 Flex Real-Time PCR System*. Available online: https://www.thermofisher.com/au/en/home/life-science/pcr/real-time-pcr/real-time-pcr-instruments/quantstudio-7-flex-real-time-pcr-system.html?icid=QuantStudioqPCRFamily-QuantStudio7flex-100915-IPAC (accessed on 23 January 2020).
31. Duffy, V.B.; Hayes, J.E.; Davidson, A.C.; Kidd, J.R.; Kidd, K.K.; Bartoshuk, L.M. Vegetable intake in college-aged adults Is explained by oral sensory phenotypes and TAS2R38 genotype. *Chemosens. Percept.* **2010**, *3*, 137–148. [CrossRef]
32. Xyris Software. FoodWorks, Brisbane, QLD, Australia. 2019. Available online: https://xyris.com.au (accessed on 18 February 2020).
33. McNaughton, S.A.; Ball, K.; Crawford, D.; Mishra, G.D. An index of diet and eating patterns is a valid measure of diet quality in an Australian population. *J. Nutr.* **2008**, *138*, 86–93. [CrossRef]
34. Collins, C.E.; Young, A.F.; Hodge, A. Diet quality is associated with higher nutrient intake and self-rated health in mid-aged women. *J. Am. Col. Nutr.* **2008**, *27*, 146–157. [CrossRef]
35. Collins, C.E.; Burrows, T.L.; Rollo, M.E.; Boggess, M.M.; Watson, J.F.; Guest, M.; Duncanson, K.; Pezdirc, K.; Hutchesson, M.J. The comparative validity and reproducibility of a diet quality index for adults: The Australian Recommended Food Score. *Nutrients* **2015**, *7*, 785–798. [CrossRef]
36. Australian Institute of Health and Welfare. *Australian Diet Quality index Project*. Available online: https://www.aihw.gov.au/getmedia/2b8340a0-b694-4eb6-b5c2-21419e06fa49/adqip.pdf.aspx?inline=true (accessed on 27 January 2020).
37. Wiltgren, A.R.; Booth, A.O.; Kaur, G.; Cicerale, S.; Lacy, K.E.; Thorpe, M.G.; Keast, R.S.; Riddell, L.J. Micronutrient supplement use and diet quality in university students. *Nutrients* **2015**, *7*, 1094–1107. [CrossRef]
38. National Health and Medical Research Council. *Eat for Health. Australian Dietary Guidelines*; NHMRC: Canberra, Australia, 2013. Available online: https://eatforhealth.govcms.gov.au/sites/default/files/content/n55_australian_dietary_guidelines.pdf (accessed on 16 February 2020).
39. Viñuela, A.; Brown, A.A.; Buil, A.; Tsai, P.C.; Davies, M.N.; Bell, J.T.; Dermitzakis, E.T.; Spector, T.D.; Small, K.S. Age-dependent changes in mean and variance of gene expression across tissues in a twin cohort. *Hum. Mol. Gen.* **2018**, *27*, 732–741. [CrossRef]

40. Lu, P.; Zhang, C.-H.; Lifshitz, L.M.; ZhuGe, R. Extraoral bitter taste receptors in health and disease. *J. Gen. Physiol.* **2017**, *149*, 181–197. [CrossRef]
41. Aburto, N.J.; Ziolkovska, A.; Hooper, L.; Elliott, P.; Cappuccio, F.P.; Meerpohl, J.J. Effect of lower sodium intake on health: Systematic review and meta-analyses. *BMJ* **2013**, *346*, f1326. [CrossRef]
42. Australian Bureau of Statistics. *National Health Survey: First Results 2017-18*. Available online: https://www.ausstats.abs.gov.au/ausstats/subscriber.nsf/0/4B3976684C09F43FCA258399001CE630/$File/4364.0.55.001%20-%20national%20health%20survey,%20first%20results,%202017-18.pdf (accessed on 2 February 2020).
43. Australian Bureau of Statistics. *National Nutrition and Physical Activity Survey*; ABS: Canberra, ACT, Australia, 2011–2012. Available online: https://www.abs.gov.au/ausstats/abs@.nsf/mf/4364.0.55.007 (accessed on 18 February 2020).
44. Barragán, R.; Coltell, O.; Portolés, O.; Asensio, E.M.; Sorlí, J.V.; Ortega-Azorín, C.; González, J.I.; Sáiz, C.; Fernández-Carrión, R.; Ordovas, J.M.; et al. Bitter, sweet, salty, sour and umami taste perception decreases with age: Sex-specific analysis, modulation by genetic variants and taste-preference associations in 18 to 80 year-old subjects. *Nutrients* **2018**, *10*, 1539. [CrossRef]

nutrients

MDPI

Article

The Association between Salt Taste Perception, Mediterranean Diet and Metabolic Syndrome: A Cross-Sectional Study

Nikolina Nika Veček [1], Lana Mucalo [2], Ružica Dragun [1], Tanja Miličević [3], Ajka Pribisalić [1], Inga Patarčić [4], Caroline Hayward [5], Ozren Polašek [6] and Ivana Kolčić [6,*]

1. University of Split School of Medicine, Šoltanska 2, 21 000 Split, Croatia; veceknika@gmail.com (N.N.V.); ruzica.dragun@yahoo.com (R.D.); ajka.relja@mefst.hr (A.P.)
2. Department of Pediatrics, Medical College of Wisconsin, Milwaukee, WI 53226, USA; lmucalo@mcw.edu
3. Department of Endocrinology and Diabetology, University Hospital Center Split, Šoltanska 1, 21 000 Split, Croatia; tanja.milicevic2@gmail.com
4. Bioinformatics and Omics Data Science Platform, Berlin Institute for Medical Systems Biology, Max-Delbrück Center for Molecular Medicine, 13092 Berlin, Germany; inga.patarcic@mdc-berlin.de
5. MRC Human Genetics Unit, IGMM, University of Edinburgh, Edinburgh EH4 2XU, UK; caroline.hayward@igmm.ed.ac.uk
6. Department of Public Health, University of Split School of Medicine, Šoltanska 2, 21 000 Split, Croatia; opolasek@mefst.hr

* Correspondence: ikolcic@mefst.hr

Received: 27 March 2020; Accepted: 20 April 2020; Published: 22 April 2020

Abstract: Metabolic syndrome (MetS) is a widespread disorder and an important public health challenge. The purpose of this study was to identify the association between salt taste perception, Mediterranean diet and MetS. This cross-sectional study included 2798 subjects from the general population of Dalmatia, Croatia. MetS was determined using the Joint Interim Statement definition, and Mediterranean diet compliance was estimated using Mediterranean Diet Serving Score. Salt taste perception was assessed by threshold and suprathreshold testing (intensity and hedonic perception). Logistic regression was used in the analysis, adjusting for important confounding factors. As many as 44% of subjects had MetS, with elevated waist circumference as the most common component (77%). Higher salt taste sensitivity (lower threshold) was associated with several positive outcomes: lower odds of MetS (OR = 0.69; 95% CI 0.52–0.92), lower odds for elevated waist circumference (0.47; 0.27–0.82), elevated fasting glucose or diabetes (0.65; 0.45–0.94), and reduced HDL cholesterol (0.59; 0.42–0.84), compared to the higher threshold group. Subjects with lower salt taste threshold were more likely to consume more fruit, and less likely to adhere to olive oil and white meat guidelines, but without a difference in the overall Mediterranean diet compliance. Salt taste intensity perception was not associated with any of the investigated outcomes, while salty solution liking was associated with MetS (OR = 1.85, CI 95% 1.02–3.35). This study identified an association between salt taste perception and MetS and gave a new insight into taste perception, nutrition, and possible health outcomes.

Keywords: salt taste perception; taste threshold; sodium chloride; metabolic syndrome; Mediterranean diet

1. Introduction

Metabolic syndrome (MetS) is a cluster of synergistic risk factors, such as abdominal obesity, arterial hypertension, hyperglycemia, and dyslipidemia that contribute to cardiovascular disease

(CVD) and mortality. There is a surge in the prevalence of all of the components of MetS, causing a worldwide pandemic and implicating both clinical and public health [1]. Given that there are still several definitions in use, which differ in their cut-off values for MetS components, the prevalence of MetS in the literature ranges anywhere between 10% and 84%, depending both on the characteristics of the sample and definition used [2]. A majority of the studies indicate that 15% to 40% of the adult population in most countries can be characterized as having MetS [3–7]. Mediterranean countries also exhibit high MetS prevalence, ranging from one quarter to one third of the population [8–10]. Unfortunately, Croatia is not at all an exception to this epidemiological situation. Previous studies have shown rather high burden of MetS in Croatian population, with crude prevalence ranging from 36% to as high as 60% in the Mediterranean region of the country [11], and 40% in the continental Croatia [12].

The main driving force contributing to such high prevalence of MetS is increase in obesity due to overconsumption of calorie-dense foods and drinks, with simultaneous decrease in physical activity levels and an alarming proportion of sedentary lifestyle [13]. The unprecedented increase in obesity worldwide has resulted from the perfect storm conditions enabled by industrial production of highly processed food, drift from the traditional food consumption practices, overall labor-saving technological advances, environmental, socio-economic and demographic changes. According to the global survey on obesity in 195 countries, 604 million adults and 108 million children were obese in 2015 [14]. Since 1980, prevalence of obesity doubled in 73 countries and increased in most other countries [14]. The Mediterranean region displays a particularly worrisome trend in childhood obesity. A recent study showed an increase in the prevalence of overweight and obesity from 22.9% in 1999 to 25.0% in 2016 among children aged 2 to 13 years in the Mediterranean part of Europe [15]. One of the explanations for this trend is departure from the traditional lifestyle and Mediterranean diet, especially in younger people from Mediterranean countries [16,17]. Interestingly, it was found that change in the food supply in the Mediterranean area, especially more readily available mass-produced food from the long supply chain (opposite from the local food markets) was associated with MetS [18]. These trends are very misfortunate and represent a double missed opportunity, because Mediterranean diet was shown to have the capacity for preventing the development of metabolic syndrome, as well as the ability to reverse it in people with or without type 2 diabetes [19,20].

Along with economic, social and environmental factors, taste perception is a major determinant of dietary choices and its impact on obesity has been previously studied. However, study results on this topic are still contradictory and inconclusive. Obese adults were reported to consume more salty foods and to have reduced salt sensitivity and higher salt preference [21–23]. Additionally, obese women showed decrease in both olfactory and taste capacity, including salt taste, compared to normal weight women [24]. On the other hand, adolescents with early onset and severe obesity displayed lower recognition thresholds, indicating higher acuity, and higher sensitivity for both sucrose and salt compared to the non-obese adolescents [25]. Finally, some studies found no association between body composition and salt taste sensitivity threshold [26], and no association between obesity and salt liking [27]. On the contrary, fat-liking was found to be associated with an increased risk of obesity [27]. Moreover, animal models have shown that high-fat diet resulted in obesity and pronounced loss of taste buds, indicating that taste loss could be a metabolic consequence of the obese state [28].

The exact mechanism on how greater sodium consumption could contribute to higher body weight remains unclear. Some authors propose that sodium intake is often accompanied by higher consumption of energy-dense foods and soft drinks [29]. The well-recognized link between high dietary salt intake, arterial hypertension and endothelial dysfunction brought the salt taste preference into focus [30]. However, studies investigating the association between salt taste and MetS are very scarce in the literature [31,32]. Furthermore, an even greater paucity exists at the intersection of salt taste sensitivity, nutrition and MetS research. Such studies would provide valuable information about the factors contributing to the MetS, which is very important for targeted prevention and treatment approaches. Prevention of MetS should be strongly emphasized since numerous studies have demonstrated that people with MetS have a 5-fold increase in risk of type 2 diabetes and a 3-fold increase in risk of CVD and related mortality, as well as increased risk for cancer [2,33]. Given these serious health consequences of MetS, any additional insight that illuminates its contributing factors is advantageous and welcome.

The aim of this study was to examine the association between salt taste threshold and suprathreshold perception and MetS components in the general population of Dalmatia, Croatia. Additionally, we examined the adherence to the Mediterranean diet according to the salt taste perception.

2. Materials and Methods

2.1. Study Participants

This cross-sectional study included subjects from the "10,001 Dalmatians" project [34], which was previously described in details [11]. For the purposes of this study, a sub-sample of 2798 subjects was used from three Dalmatian settlements: the Island of Vis (n = 390, sampled in 2011), the Island of Korcula (n = 1908, sampled during 2012–2016 period) and the City of Split (n = 500; sampled in 2012–2013). A population-based sample was gathered via direct postal invitations, radio announcements and support from the local general practitioners and local government officials. The only exclusion criterion was age of <18 years old. All potential subjects were informed on the study aims and goals, expected benefits and risks, after which those who decided to participate have signed the informed consent. The study was conducted in accordance with the Declaration of Helsinki, and approved by the Ethical Board of the University of Split School of Medicine (2181-198-03-04/10-11-0008).

2.2. Data Collection and Measurements

Each subject provided a fasting blood sample, filled a self-administered questionnaire and had blood pressure and anthropometric measurements done. Trained medical professionals (medical doctor or research nurse) collected data on medical history and previous diagnoses, as well as on medications being used for: hypertension, coronary heart disease (CHD), cerebrovascular insult (CVI), type 2 diabetes, hyperlipidemia, cancer, bipolar disorder and gout.

The questionnaire included questions on demographic characteristics (age and gender), socioeconomic status (education and material status), smoking, alcohol consumption, physical activity, and dietary habits. Education was classified in three categories based on the number of years of completed schooling (primary, secondary, and university level). Material status was based on the subjects' material possessions and classified into quartiles, as described previously [11]. Namely, subjects answered to 16 questions about their material possessions (heating system, wooden floors, video/DVD recorder, telephone, computer, two TVs, freezer, dishwasher, water supply system, flushing toilet, bathroom, library with more than 100 books, paintings or other art objects, a car, vacation house or second apartment, boat). Based on the sum of all positive responses, subjects were divided into quartiles.

Based on smoking habits, we divided subjects in active smokers (for whom we calculated the number of pack-years, by multiplying the number of cigarettes smoked per day with the number of years they smoked), ex-smokers (stopped smoking more than 1 year ago) and non-smokers. Alcohol consumption was measured as the total number of units of alcohol ingested per week, while physical activity was self-assessed as light, moderate or intensive.

The Mediterranean Diet Serving Score (MDSS) was calculated as suggested by Monteagudo et al., based on the food frequency questionnaire consisting of 55 questions [35]. Shortly, this scoring approach requires daily consumption of cereals, vegetables, fruit and olive oil (in each main meal), one or two daily servings of nuts and dairy products, daily moderate alcohol intake (ideally a glass of wine per day), consumption of fish and legumes several times per week, while other meats and sweets should be consumed rarely, once or twice per week [35]. The maximum MDSS value is 24 points, and the cut-off value of ≥14 points was used to define compliance with the principles of the Mediterranean diet [11]. Additionally, we asked subjects about their habits of adding salt before tasting food, and they could have responded as never, occasionally, often or almost always.

Blood pressure was measured in a sitting position after a rest period of at least ten minutes. We measured the blood pressure twice in each subject in order to calculate the average value, which was used in the analysis.

Anthropometric measures included body height and mass, measured using a combined stadiometer and a scale instrument (model 704, Seca GmBH & Co., Hamburg, Germany), while waist and hip circumferences were measured in millimeters using an inelastic measuring tape. Using these measures we have calculated body mass index (BMI), waist-to-hip ratio (WHR) and waist-to-height ratio (WHtR), as relative estimates of central obesity. During the anthropometric measurement, subjects were dressed in underwear or light clothing.

2.3. Biochemistry Measurements and Metabolic Syndrome Definition

After blood collection, the sample was processed in a field laboratory and stored in a −80 °C freezer. Biochemical analysis was performed at accredited Brayer Polyclinic laboratory in Zagreb using standard methods for determining biochemical parameters. In this study, we used data on fasting glucose (mmol/L), triglycerides (mmol/L), total cholesterol (mmol/L), LDL cholesterol (mmol/L), HDL cholesterol (mmol/L) and HbA1c (%).

Metabolic syndrome was defined according to the Joint Interim Statement definition [36]. The subject had to have at least three of the following criteria: elevated waist circumference (≥80 cm for women and ≥94 cm for men), elevated triglycerides (≥1.7 mmol/L) or using medications, reduced HDL concentration (<1.0 mmol/L in men, <1.3 mmol/L in women) or using medications, elevated systolic and/or diastolic blood pressure (≥130/85 mmHg) or using medications for hypertension, elevated fasting glucose (≥5.6 mmol/L) or using medications for diabetes [36].

2.4. Measurement of Salt Taste Perception

Salt taste perception was assessed by threshold and suprathreshold testing designed according to the ISO standards [37], and performed by trained researchers. Subjects had to restrain from chewing gum, smoking, eating and drinking anything except water, at least half an hour before testing. All of the tests were performed using water solutions of table salt. Solutions were prepared daily and kept at room temperature. Due to the conditions of the field testing away from the laboratory, we used a standardized commercial mineral water, with following content: 64.2 mg/L of Ca^{2+}, 32.1 mg/L of Mg^{2+}, 1.7 mg/L of Na^{+}, and 2.8 mg/L of Cl^{-}.

For salt taste recognition threshold, we used five concentrations, starting with the weakest solution and equally increasing concentrations (0.22 log increment). These solutions, each in an individual volume of 10 mL, were presented to the subjects starting from the lowest concentration of 8.21 mmol/L (0.48 g of NaCl dissolved in 1 L of water), followed by 13.69 mmol/L, 22.81 mmol/L, 38.02 mmol/L, while the highest concentration was 63.37 mmol/L. We performed a pilot testing (n = 32) using these concentrations in order to confirm that they are appropriate for use in the general population across different ages, identifying both people who recognize the lowest concentration and those not recognizing the highest concentration.

Subjects were blinded to the taste quality presented to them and increasing concentrations of solutions were used until the point when they correctly recognized salty taste. Subjects were instructed to taste the solution for a couple of seconds and they were allowed to swallow the solution before providing their answer. The correct answer was denoted as the ordinal number of the solution, starting with the number 1, which marked the lowest concentration and number 5 marked the highest concentration, while number 6 was used in cases when subjects didn't recognize salty taste of the solution with the highest concentration. Between testing solution presentations, subjects were instructed to rinse their mouth with the same water used for preparation of testing solutions. Subjects were presented with only one solution at a time in order to make the testing procedure overall less time consuming and less cumbersome, especially for elderly subjects. With each solution presentation, subjects were asked whether the solution tasted like plane water or something else, and in case of confirmatory answer, they were asked to identify the quality of the taste (sweet, salty, sour or bitter). Because of this simplified testing procedure, unlike the usual 3-alternative forced choice or 2-alternative forced choice, each subject performed two recognition threshold testing rounds, with a break of at least of 30 min in between them. Based on these two testing responses, a geometric mean was calculated using the ordinal numbers of correctly recognized solutions. Those subjects who had a geometric mean result of ≤2.0 were considered as having a lower salt taste recognition threshold, which corresponds to the higher sensitivity and acuity. A geometric mean between 2.1 and 4.0 was considered as intermediate threshold, while those subjects with a result of ≥4.1 and those who didn't recognize the highest salt concentration at both testing times were considered as having a higher taste threshold and lower salt taste acuity.

After threshold testing, suprathreshold salt taste perception was tested as perceived intensity and hedonic response (liking), using a 10 mL of table salt solution with the concentration of 137 mmol/L (8 grams of NaCl per liter). This concentration is slightly more than double the highest threshold solution concentration (3.7 g/L), and it was used before as the highest concentration for threshold testing [38], pointing to be a possible concentration for suprathreshold testing, but not too concentrated to be off-putting. Suprathreshold measurements were available in a subsample of the subjects consisting of 1155 people sampled after 2012 (926 subjects from Korcula and 229 subjects from Split subsample). The Labeled Magnitude Scale (LMS) was used to estimate the taste intensity perception [39]. In short, we placed words describing the intensity of the salty taste along a vertical line, without any numeric markers. The words "no sensation" were placed at the start of the line (0 mm), "barely detectable" at 2 mm, "weak sensation" at 7 mm, "moderate" at 20 mm, "strong" at 40 mm, "very strong" at 61 mm, and "strongest imaginable" was placed at 114 mm from the beginning of the line [37]. Subjects have practiced using the scale in at least one tasting attempt, after which they placed their final mark on the LMS line immediately after tasting the solution. Their intensity response was measured in millimeters as the distance of the subjects' mark relative to the beginning of the scale. Based on the corresponding wording along the line, we have divided subjects into three groups in order to simplify the analysis. Subjects who marked their response between 0 and 39 mm on the LMS line were considered as having felt nothing to medium strong intensity (lower perceived intensity). Subjects responding between 40 to 61 mm found the solution to be strong to very strong, and those who responded between 62 and 114 mm thought that the solution was extremely strong in intensity.

Hedonic perception (liking) was tested using the same suprathreshold salt solution concentration, using the Labeled Affective Magnitude scale (LAM) [40]. LAM scale is also a vertical line with a total length of 100 mm, where semantic labels "greatest imaginable dislike", "neither like nor dislike", and "greatest imaginable like" were placed at 0, 50 and 100 mm, respectively, but without displayed numbers [40]. Subjects were asked to make a mark indicating how much they liked or disliked the taste of the concentrated salt solution. Their responses were turned into a numeric variables in the way that the middle of the scale was regarded as a 0 mm, and negative responses were below that point (the start of the line was marked as −50 mm and it indicated "greatest imaginable dislike"), while positive response were above the middle point (up to +50 mm, indicating "greatest imaginable like"). Based on these distances, we have divided subjects into three groups; those who disliked the solution (response between −50 to −10 mm), those who neither liked nor disliked it (−9 to +9 mm), and those who liked concentrated salty solution (+10 to +50 mm).

2.5. Statistical Analysis

Categorical variables were presented as absolute numbers and percentages. Numerical variables were tested for normality using Kolmogorov–Smirnov test and central tendency was presented using medians and interquartile ranges (IQR), due to mostly non-normal data distribution. Differences between groups were tested using chi-square test for categorical variables, and Mann–Whitney U test or Kruskal–Wallis tests were used for numerical variables, depending on the number of groups. Spearman correlation test was used to identify bivariate correlation between age, threshold and suprathreshold salt taste perception, and the frequency of adding salt before tasting the food.

The association between the presence of the MetS and salt taste threshold (full sample, $n = 2798$), salt taste intensity and hedonic perception (subsample, $n = 1155$) was tested by multivariate logistic regression analysis. Several logistic regression models were created and adjusted for confounders. Prevalent MetS and each of the five MetS components were considered to be dependent variables. The association between independent variables (taste threshold, taste intensity and taste hedonic perception) and dependent variables was adjusted for age, gender, place of residence (Island Vis, Island Korcula, City of Split), education level (primary, secondary, university), quartiles of material status, BMI (in all regression models except for elevated waist circumference), smoking (never-smokers, ex-smokers, active smokers), alcohol intake (units/week), physical activity (low, moderate, intensive), Mediterranean Diet Serving Score (MDSS), and adding salt before tasting food (never, occasionally, often, almost always).

Additionally, the association between the Mediterranean diet (MDSS ≥14 points) and it's components (dependent variables) and salt taste perception (three independent variables) was tested using multivariate logistic regression analysis. Models were adjusted to the same confounding variables, but excluding the MDSS score.

Significance level was set at $p < 0.05$. Data analysis was conducted using IBM SPSS Statistics for Windows, v21.0 (IBM, Armonk, NY, USA).

3. Results

This cross-sectional study included 2798 subjects from the Island of Vis, the Island of Korcula and the City of Split. According to their salt taste recognition threshold, we divided subjects into three categories, where lower taste threshold indicated higher taste acuity. Subjects with higher salt taste threshold were on average older, had higher proportion of men, lower education level, higher anthropometric indices, and highest average values for all of the MetS constituent components, except for HDL cholesterol (Table 1). There were no differences in habits, except in the Mediterranean diet adherence and adding salt before tasting food. Subjects with higher salt threshold added salt to their food more frequently compared to subjects with both lower and intermediate threshold (Table 1).

Table 1. Subjects' characteristics according to the salt taste recognition threshold perception.

	Lower Salt Taste Threshold/Higher Acuity $n = 1094$	Intermediate Salt Taste Threshold/Acuity $n = 1236$	Higher Salt Taste Threshold/Lower Acuity $n = 468$	p
	Socio-demographic characteristics			
Age (years); median (IQR)	52.0 (23.0)	56.0 (21.0)	59.0 (19.0)	<0.001 *
	Gender; n (%)			
Women	751 (68.6)	746 (60.4)	260 (55.6)	<0.001 †
Men	343 (31.4)	490 (39.6)	208 (44.4)	
	Place of residence; n (%)			
Vis	123 (11.2)	110 (8.9)	157 (33.5)	<0.001 †
Korčula	727 (66.5)	910 (73.6)	271 (57.9)	
Split	244 (22.3)	216 (17.5)	40 (8.5)	
	Education level; n (%)			
Primary school	180 (16.6)	294 (24.1)	175 (37.8)	<0.001 †
Secondary school	588 (54.1)	633 (51.9)	210 (45.4)	

Table 1. *Cont.*

	Lower Salt Taste Threshold/Higher Acuity $n = 1094$	Intermediate Salt Taste Threshold/Acuity $n = 1236$	Higher Salt Taste Threshold/Lower Acuity $n = 468$	p
University level	318 (29.3)	293 (24.0)	78 (16.8)	
		Anthropometry		
Weight (kg); median (IQR)	75.9 (21.0)	78.0 (21.2)	79.0 (18.5)	0.001 *
BMI (kg/m^2); median (IQR)	25.5 (5.6)	26.1 (5.6)	26.2 (5.7)	<0.001 *
WHR; median (IQR)	0.89 (0.11)	0.92 (0.11)	0.94 (0.11)	<0.001 *
WHtR; median (IQR)	0.54 (0.09)	0.55 (0.09)	0.57 (0.10)	<0.001 *
		Metabolic syndrome components		
Waist circumference (mm); median (IQR)	919.5 (158.0)	942.5 (150.0)	967.5 (132.3)	<0.001 *
Fasting glucose (mmol/L); median (IQR)	5.2 (0.9)	5.4 (1.0)	5.5 (1.2)	<0.001 *
Systolic blood pressure; median (IQR)	125.0 (25.0)	130.0 (20.0)	135.0 (22.6)	<0.001 *
Diastolic blood pressure; median (IQR)	80.0 (15.0)	80.0 (10.0)	80.0 (11.6)	<0.001 *
Triglycerides (mmol/L); median (IQR)	1.1 (0.7)	1.2 (0.8)	1.3 (0.9)	<0.001 *
HDL (mmol/L); median (IQR)	1.5 (0.5)	1.4 (0.5)	1.4 (0.5)	0.351 *
		Other biochemical parameters		
Total cholesterol (mmol/L); median (IQR)	5.8 (1.6)	5.8 (1.6)	5.9 (1.5)	0.284 *
LDL cholesterol (mmol/L); median (IQR)	3.7 (1.5)	3.7 (1.3)	3.7 (1.5)	0.911 *
HbA1c (%); median (IQR)	5.3 (0.6)	5.3 (0.5)	5.3 (0.6)	0.068 *
		Habits		
Smoking; n (%)				
Never-smokers	519 (47.7)	602 (49.2)	225 (48.6)	0.833 †
Ex-smokers	267 (24.5)	275 (22.5)	111 (24.0)	
Active smokers	302 (27.8)	346 (28.3)	127 (27.4)	
Pack-years in smokers; median (IQR)	12.0 (18.0)	14.0 (24.0)	20.0 (30.0)	<0.001 *
Alcohol intake (units/week); median (IQR)	6.8 (18.9)	6.7 (20.3)	6.8 (20.3)	0.087 *
Physical activity; n (%)				
Low	262 (24.3)	263 (21.6)	103 (22.6)	0.528 †
Moderate	717 (66.5)	834 (68.6)	303 (66.4)	
Intensive	100 (9.3)	119 (9.8)	50 (11.0)	
Mediterranean diet adherence (MDSS points); median (IQR)	11.0 (6.0)	10.0 (5.0)	11.0 (6.0)	0.011 *
		Adding salt before tasting food; n (%)		
Never	430 (42.4)	504 (43.6)	139 (31.0)	<0.001 †
Occasionally	250 (24.6)	316 (27.3)	102 (22.7)	
Often	256 (25.2)	251 (21.7)	164 (36.5)	
Almost always	79 (7.8)	86 (7.4)	44 (9.8)	

IQR—interquartile range; BMI—body mass index; WHR—waist-to-hip ratio; WHtR—waist-to-height ratio; MDSS—Mediterranean Diet Serving Score; * Kruskal-Wallis test; † chi-square test.

Subjects with MetS were on average older, had higher proportion of men and Vis Island inhabitants, lower education level, higher anthropometric indices, and higher average values for all of the MetS constituent components, as well as other biochemical parameters (Appendix A, Table A1). Subjects with MetS were less frequently active smokers, but with higher average pack-years among smokers, had slightly higher proportion of subjects with intensive level of physical activity, greater average score of the Mediterranean diet adherence, and they added salt to their food more frequently compared to subjects without MetS (Appendix A, Table A1).

Salt taste threshold was correlated negatively with intensity perception and positively with age and the habit of adding salt before food tasting (all $p < 0.001$). Additionally, salt taste intensity perception was correlated negatively with age and hedonic perception (both $p < 0.001$) (Table 2).

Table 2. Correlation between age, salt adding habit, threshold and suprathreshold salt taste perception, data presented are Spearman's rho correlation coefficients (*P* values).

	Salt Taste Threshold	Salt Taste Intensity Perception	Salt Taste Hedonic Perception	Adding Salt Before Food Tasting
Age	0.224 (<0.001)	−0.170 (<0.001)	0.038 (0.198)	−0.016 (0.387)
Salt taste threshold		−0.132 (<0.001)	0.043 (0.143)	0.092 (<0.001)
Salt taste intensity perception			−0.399 (<0.001)	−0.024 (0.424)
Salt taste hedonic perception				0.066 (0.025)

We observed high prevalence of MetS, with differences according to the salt taste threshold sensitivity, where 57.9% of subjects with lower taste acuity had MetS, compared to 38.3% of subjects with higher taste acuity (lower salt taste threshold) (Table 3). Elevated waist circumference was the most common metabolic syndrome component in all threshold sensitivity groups (as high as 79.5%), followed by elevated blood pressure (up to 56.2% in subjects with higher threshold) (Table 3). The only MetS component without significant difference between threshold sensitivity groups was HDL cholesterol.

Table 3. Prevalence of metabolic syndrome components according to the salt taste recognition threshold perception, data are presented as *n* (%).

	Lower Salt Taste Threshold/Higher Acuity *n* = 1094	Intermediate Salt Taste Threshold/Acuity *n* = 1236	Higher Salt Taste Threshold/Lower Acuity *n* = 468	p †
Elevated waist circumference	797 (73.5)	980 (79.5)	367 (79.4)	0.001
Elevated glucose or diabetes present	90 (8.4)	131 (10.8)	80 (17.4)	<0.001
Elevated blood pressure or hypertension present	423 (38.7)	580 (46.9)	263 (56.2)	<0.001
Elevated triglycerides	250 (22.9)	340 (27.5)	143 (30.6)	0.002
Reduced HDL	207 (18.9)	239 (19.3)	86 (18.4)	0.899
Metabolic syndrome present	419 (38.3)	564 (45.6)	271 (57.9)	<0.001

† chi-square test.

After stratification according to the age, only middle-aged subjects (35–65 years old) presented with higher salt taste threshold more frequently in subjects with MetS (18.6% vs. 12.0%), and less frequently with lower threshold compared to the subjects without MetS (38.6% vs. 43.2%). Similar results were present in the subgroup of subjects older than 65 years, but with borderline insignificant result ($p = 0.056$) (Appendix A, Table A2). There were no differences in either salt taste intensity or hedonic perception between subjects with MetS and those without it, in any of the age groups (Appendix A, Table A2).

There was a borderline insignificant result in the Mediterranean diet compliance between subjects with different taste threshold perception (Appendix A, Table A3). We also observed that subjects with higher salt taste thresholds more frequently complied with the Mediterranean pyramid recommendations for olive oil, legumes, fish, and white meat, while they less frequently complied with fruit and potatoes guidelines, compared to subjects in lower taste threshold group (Appendix A, Table A3). Similar results were obtained in the regression analysis adjusted for confounding factors, where subjects with lower salt taste threshold were more likely to consume fruit several times a day (OR = 1.52, 95% CI 1.16–1.97; p = 0.002), the same as subjects with intermediate threshold (OR = 1.41, 95% CI 1.09–1.81; p = 0.008), compared to subjects with higher threshold (Table 4). The opposite was recorded for olive oil and white meat consumption, while the result for fish was borderline insignificant (OR = 0.76, 95% CI 0.58–1.00; p = 0.053). There were no differences in consumption of vegetables, legumes, red meat and sweets, or in overall compliance to the Mediterranean diet between subjects with lower and higher salt taste thresholds (Table 4). There were also no differences in the Mediterranean diet or in food groups compliance with regard to the salt taste intensity nor with hedonic perception (Table 4).

Subjects with lower salt taste threshold had lower odds of having elevated waist circumference (OR = 0.47, 95% CI 0.27–0.82; p = 0.008; fully adjusted model), the same as for having elevated fasting glucose or diabetes (OR = 0.65, 95% CI 0.45–0.94; p = 0.022), compared to higher threshold group (Table 5). Subjects with both lower and intermediate threshold had lower odds of having reduced HDL cholesterol (OR = 0.59, 95% CI 0.42–0.84; p = 0.003 and OR = 0.65, CI 95% 0.47–0.91; p = 0.011, respectively) and lower odds of having MetS (OR = 0.69, 95% CI 0.52–0.92; p = 0.013 and OR = 0.75, 95% CI 0.57–0.99; p = 0.044, respectively) (Table 5).

Salt taste intensity perception did not show significant association with metabolic syndrome and its components. Only subjects who liked salty solution had higher odds of having metabolic syndrome (OR 1.85, CI 95% 1.02–3.35; p = 0.042), compared to subjects who disliked the solution (Table 5). Those subjects who liked salty solution also had a borderline insignificantly higher odds for having elevated blood pressure or diagnosis of hypertension (OR = 1.79, 95% CI 0.97–3.31; p = 0.063) and borderline insignificantly lower odds for having reduced HDL cholesterol (OR = 0.48, 95% CI 0.22–1.03; p = 0.058) (Table 5).

Pictorial presentation of the main findings of the study, with the results from adjusted logistic regression models are presented in Figure 1.

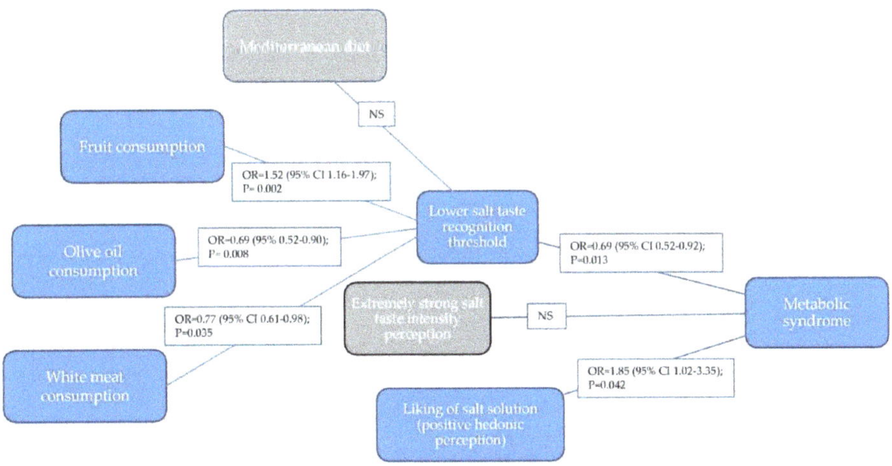

Figure 1. Pictorial presentation of the main findings of the study (results are from adjusted logistic regression model; NS—non-significant).

Table 4. Association between Mediterranean diet compliance and salt taste threshold and suprathreshold perception.

	Fruit OR (95% CI); *p*	**Vegetables** OR (95% CI); *p*	**Olive Oil** OR (95% CI); *p*	**Legumes** OR (95% CI); *p*	**Fish** OR (95% CI); *p*	**White Meat** OR (95% CI); *p*	**Red Meat** OR (95% CI); *p*	**Sweets** OR (95% CI); *p*	**MDSS Compliance** OR (95% CI); *p*
Salt taste threshold									
Higher salt taste threshold (lower acuity)	referent	referent	referent	referent	referent	referent	referent	referent	referent
Intermediate salt taste threshold	1.41 (1.09–1.81); 0.008	0.96 (0.74–1.25); 0.771	0.77 (0.58–1.00); 0.051	0.83 (0.64–1.09); 0.187	0.83 (0.62–1.10); 0.185	0.82 (0.64–1.05); 0.116	0.94 (0.71–1.23); 0.639	1.11 (0.84–1.45); 0.477	0.93 (0.69–1.24); 0.603
Lower salt taste threshold (higher acuity)	1.52 (1.16–1.97); 0.002	1.02 (0.78–1.33); 0.881	0.69 (0.52–0.90); 0.008	0.96 (0.72–1.26); 0.749	0.76 (0.58–1.00); 0.053	0.77 (0.61–0.98); 0.035	1.09 (0.82–1.45); 0.544	1.17 (0.88–1.56); 0.295	1.04 (0.77–1.41); 0.807
Salt taste intensity perception §									
No sensation to medium strong	referent	referent	referent	referent	referent	referent	referent	referent	referent
Strong to very strong	0.69 (0.43–1.11); 0.124	1.07 (0.63–1.80); 0.812	1.05 (0.65–1.70); 0.836	0.90 (0.51–1.58); 0.700	1.14 (0.71–1.83); 0.588	1.08 (0.68–1.72); 0.753	1.03 (0.61–1.75); 0.913	0.83 (0.51–1.36); 0.464	0.65 (0.37–1.15); 0.141
Extremely strong	1.08 (0.64–1.84); 0.772	1.09 (0.61–1.93); 0.776	1.23 (0.72–2.08); 0.448	1.01 (0.54–1.89); 0.985	1.25 (0.74–2.11); 0.412	1.31 (0.79–2.19); 0.299	1.24 (0.69–2.22); 0.473	0.67 (0.38–1.18); 0.164	0.80 (0.43–1.50); 0.483
Salt taste hedonic perception §									
Dislike	referent	referent	referent	referent	referent	referent	referent	referent	referent
Nor like nor dislike	1.20 (0.84–1.70); 0.320	1.03 (0.71–1.50); 0.884	0.73 (0.52–1.04); 0.081	1.32 (0.88–1.99); 0.175	1.16 (0.81–1.65); 0.430	1.14 (0.81–1.60); 0.452	1.18 (0.81–1.74); 0.383	1.14 (0.79–1.65); 0.470	0.77 (0.48–1.22); 0.263
Like	1.02 (0.61–1.71); 0.940	0.84 (0.48–1.48); 0.884	1.65 (0.95–2.88); 0.075	0.91 (0.48–1.71); 0.772	0.89 (0.54–1.48); 0.660	1.23 (0.76–2.01); 0.401	1.19 (0.68–2.08); 0.540	0.77 (0.44–1.37); 0.375	1.21 (0.66–2.21); 0.537

All models were adjusted for age, gender, place of residence, education level (three categories: primary, secondary, university), material status (quartiles), BMI, smoking (three categories: never-smokers, ex-smokers, active smokers), alcohol intake (units/week), physical activity (three categories: low, moderate, intensive), adding salt before tasting food (four categories: never, occasionally, often, almost always); § calculated from the sample subset including 1155 subjects.

Table 5. Association between metabolic syndrome and its components and salt taste threshold and suprathreshold perception.

	Elevated Waist Circumference OR (95% CI); *p*	Elevated Blood Pressure or Diagnosis of Hypertension OR (95% CI); *p*	Elevated Fasting Plasma Glucose or Diagnosis of Diabetes OR (95% CI); *p*	Elevated Triglycerides or Using Medications OR (95% CI); *p*	Reduced HDL Cholesterol or Using Medications OR (95% CI); *p*	Metabolic Syndrome OR (95% CI); *p*
Salt taste threshold						
Higher salt taste threshold (lower acuity)	referent	referent	referent	referent	referent	referent
Intermediate salt taste threshold	1.03 (0.72–1.46); 0.880	0.85 (0.65–1.13); 0.276	0.71 (0.48–1.06); 0.097	0.89 (0.68–1.19); 0.453	0.65 (0.47–0.91); 0.011	0.75 (0.57–0.99); 0.044
Lower salt taste threshold (higher acuity)	0.47 (0.27–0.82); 0.008	0.89 (0.66–1.20); 0.433	0.65 (0.45–0.94); 0.022	0.76 (0.57–1.02); 0.070	0.59 (0.42–0.84); 0.003	0.69 (0.52–0.92); 0.013
Salt taste intensity perception §						
No sensation to medium strong	referent	referent	referent	referent	referent	referent
Strong to very strong	1.17 (0.64–2.17); 0.608	1.57 (0.85–2.77); 0.123	0.97 (0.45–2.11); 0.941	0.89 (0.52–1.52); 0.660	1.11 (059–2.08); 0.743	1.20 (0.81–1.77); 0.362
Extremely strong	0.91 (0.47–1.75); 0.766	1.18 (0.63–2.23); 0.602	0.90 (0.36–2.21); 0.812	0.87 (0.48–1.59); 0.648	1.23 (0.63–2.41); 0.550	0.65 (0.35–1.22); 0.177
Salt taste hedonic perception §						
Dislike	referent	referent	referent	referent	referent	referent
Neither like nor dislike	1.04 (0.66–1.63); 0.883	1.13 (0.75–1.70); 0.554	0.74 (0.39–1.41); 0.360	0.78 (0.50–1.20); 0.254	1.01 (0.65–1.56); 0.976	1.35 (0.91–2.01); 0.139
Like	0.64 (0.35–1.19); 0.159	1.79 (0.97–3.31); 0.063	0.47 (0.16–1.33); 0.154	1.45 (0.82–2.59); 0.197	0.48 (0.22–1.03); 0.058	1.85 (1.02–3.35); 0.042

All models were adjusted for age, gender, place of residence, education level (three categories: primary, secondary, university), material status (quartiles), BMI (in all regression models except for elevated waist circumference), smoking (three categories: never-smokers, ex-smokers, active smokers), alcohol intake (units/week), physical activity (three categories: low, moderate, intensive), Mediterranean Diet Serving Score (MDSS), adding salt before tasting food (four categories: never, occasionally, often, almost always); § calculated from the sample subset including 1155 subjects.

4. Discussion

The most important and new findings of this study include identification of the lower odds for MetS and most of the MetS components in subjects with higher salt taste sensitivity (lower threshold), as well as the association of salt taste threshold with several Mediterranean diet food groups in the general population. Contrary to this, we found no indication of the association between suprathreshold salt taste perception and these outcomes, except for higher odds of MetS in subjects with higher liking of salty solution.

MetS is a common disorder in the general population [41]. The same situation is present in Croatia. As many as 44% of subjects included in this study had MetS, which is similar to previously reported MetS prevalence in both adult population in Croatia [42,43] and in obese children and adolescents [44], while some studies found crude MetS prevalence to be even greater than 55% [45]. Some of the differences between these studies can be explained by different diagnostic criteria being used and different population groups included. The MetS is associated with many adverse outcomes, such as increased risk of cardiovascular disease, diabetes, chronic kidney disease and total mortality [46,47]. This makes MetS a very important public health challenge and a research target, in order to identify risk factors behind its development and useful approaches in prevention and treatment. Many of the MetS risk factors have been identified and repeatedly confirmed, such as poor nutrition and lack of physical activity. Determinants influencing these risk factors are now becoming increasingly important, and taste perception is surely among them [32,48]. Taste and olfaction form the basis of flavor perception, and as such, they are well-recognized and major predictors of food choices, dietary patterns, body composition and consequent health outcomes [49]. The sense of taste has been extensively studied and many determinants of individual differences in taste perception have been identified, such as genetic factors, age, habits and lifestyle factors, alongside with various pathologies and metabolic diseases, such as obesity [50]. Salt taste perception was investigated to a lesser extent, and most commonly in relation to the salt sensitivity (change in blood pressure depending on the change in salt intake), hypertension and salt intake [51,52]. The association between salt intake and hypertension was indeed extensively studied [30]. Salt taste perception was substantially less frequently investigated, especially in association with the MetS. There are only a handful of studies published so far on this topic [31]. Hence, we aimed to fill this gap and examine the association of both salt taste threshold sensitivity and suprathreshold perception with different health outcomes included in the MetS definition. Namely, based on the regression analysis, we found that subjects with lower salt taste threshold, indicating higher salt taste sensitivity, had 31% lower odds of having MetS (OR = 0.69; 95% CI 0.52–0.92). They also had 53% lower odds for elevated waist circumference (OR = 0.47; 95% CI 0.27–0.82), 35% lower odds for having elevated fasting glucose or diabetes diagnosis (OR = 0.65; 95% CI 0.45–0.94), and 41% lower odds for having reduced HDL cholesterol (OR = 0.59; 95% CI 0.42–0.84), compared to the higher threshold group, while the result for elevated triglycerides was borderline insignificant (OR = 0.76; 95% CI 0.57–1.02). These results confirm previous results of increased salt taste threshold in subjects with MetS compared to the subjects without MetS, which was independent of sex, age and BMI [31]. Another study found a positive association between sodium excretion, indicating higher intake, and components of MetS, such as blood pressure, waist circumference, triglycerides, and fasting glucose and an inverse association with HDL cholesterol [53]. Additionally, subjects with higher sodium excretion also had a higher body fat percentage, body fat mass, and insulin levels, pointing to the high-salt diet as a significant risk factor for MetS [53]. Additionally, several studies have identified the association between salt taste perception and obesity, which is a fundamental MetS component. For instance, one of these studies, including only healthy adults, showed that salt taste threshold was higher in people with higher BMI, with a similar result for olfactory threshold, indicating that increasing BMI was associated with a decrease in olfactory and taste sensitivity [54]. A decreased taste capacity was found with increase in visceral fat, with a negative correlation between salt taste threshold and BMI, total fat mass and visceral fat, as well as with insulin, leptin, glucose, and HDL cholesterol in healthy women [24]. However, there are studies that showed the opposite results. For example, Hardikar et al. found that obese subjects had lower thresholds for sucrose and salt, as well as higher ratings of intensity, indicating a higher sensitivity to sweet and salty tastes, compared to lean subjects [55]. Donaldson et al. pointed that threshold was lower for salt, unchanged for sweet and higher for bitter and sour taste in obese adults [22].

We found an association between higher salt taste threshold and elevated blood pressure and/or previous diagnosis of hypertension in bivariate analysis, but this association was not confirmed in multivariate analysis. This was actually the only MetS component not showing the association with salt taste threshold in our subjects, while controlling for important confounding factors. This is in contrast with previous studies, which showed that higher salt taste sensitivity threshold was associated with increased blood pressure [56], even so in women with normal-range blood pressure [57], and also in response to exercise [58]. However, some studies did not manage to demonstrate the association between salt taste threshold and hypertension [59] or between suprathreshold intensity perception and either hypertension or mean blood pressure [60]. An inverse association was reported between salt taste intensity perception and the frequency of adding salt to foods [60]. This habit of adding salt to the food before tasting was rather prevalent in our subjects, and it was positively correlated with both salt taste threshold and hedonic perception, but not with age or with salt taste intensity perception. Such habit should be strongly discouraged because daily salt consumption in food is unequivocally associated with increase in blood pressure and risk for hypertension. Moreover, animal and human studies showed that long-term intake of high-sodium diet increases the risk of obesity, insulin resistance and diabetes development, irrespective of total energy or glucose intake [61,62]. Lanaspa et al. performed an interesting study in which they elaborated potential underlying pathophysiological mechanism by which salt may cause obesity and MetS [63]. They showed that prolonged high-salt diet in mice generated endogenous fructose production by activating hepatic aldose reductase (AR), what resulted in hepatic sorbitol and triglyceride accumulation, as well as serum leptin elevation [63]. Obesity, on the other hand, can be both caused and lead to hedonic eating (eating for pleasure and not for hunger), by disrupting the normal taste input processing [64]. Elevated BMI was found to be related to changes in the brain activity in regions involved in salt taste perception [23]. Indeed, it was shown that salt taste engages various brain regions that modulate reward, taste processing, and executive control in eating [64], possibly resulting in greater salt consumption in overweight/obese people, in association with reduced salt sensitivity and a higher salt preference [23,64].

The association between salt taste perception and dietary choices and habits are not well understood or extensively studied. Only a handful of studies have so far examined this topic [65,66]. One such study showed that healthy adults who were hyposensitive to salty taste consumed more bakery and salty baked products, more saturated fat-rich products, and less soft drinks compared to people with higher taste acuity [65]. To our best knowledge, this is the first study to investigate the association between salt taste perception and adherence to the Mediterranean diet. Our results showed that subjects with lower salt taste threshold more frequently complied with the Mediterranean pyramid recommendations for fruit, but less so with the recommendations for consumption of olive oil and white meat (and borderline insignificant for fish), compared to subjects with higher salt taste threshold. After adjusting for important confounding factors, there was no difference in the overall compliance to the Mediterranean diet between lower and higher salt taste threshold group, possibly due to the opposite associations found for fruits and olive oil.

Overall, compliance with the Mediterranean diet was rather low in our subjects (23%). As we reported previously, it was particularly low in younger age groups, and lower in men compared to women [17]. Unfortunately, this departure from the traditional Mediterranean diet in the population of Dalmatia represents potentially invaluable losses in the domains of population health, environmental sustainability, local economy and cultural heritage preservation [67]. Population health might be on the line already, given that the recent generations of the Adriatic islanders have lost their advantage in life expectancy at birth compared to the mainland population, possibly due to diminishing adherence to the Mediterranean diet and traditional lifestyle [68].

Furthermore, the loss of Mediterranean diet represents an immense missed opportunity for primary prevention of CVD. Namely, it was shown that people compliant with the Mediterranean diet had a 30% risk reduction for the major cardiovascular event (myocardial infarction, stroke, or death from cardiovascular causes) [69]. Besides primary and secondary prevention of CVD, Mediterranean diet plays a role in improving health in overweight and obese patients, preventing the increase in weight and waist circumference in non-obese people, and both improving MetS and reducing its incidence [70]. A meta-analysis including 33,847 individuals showed that high adherence to the Mediterranean diet reduced the risk of MetS by 19% [71]. In addition, higher intake of some polyphenols, which have been suggested to be partly responsible for the beneficial effects of the Mediterranean diet, showed an inverse association with blood pressure, fasting plasma glucose, HDL cholesterol and triglycerides [72].

Further studies are needed to elucidate the link between salt taste perception and Mediterranean diet adherence, especially since we did not find any apparent association between salt taste intensity or hedonic perception and the Mediterranean diet compliance in our study. The only significant association we have identified between suprathreshold salt taste perception was a 85% increase in odds for MetS presence in subject who positively rated salty solution compared to those who disliked it. Additionally, we identified a suggestive higher odds (borderline insignificant) for elevated blood pressure or hypertension and lower odds for reduced HDL cholesterol in subjects who liked salty solution. Contrary to our expectations and due to the negative correlation between salt taste threshold and intensity perception, we found no significant associations with the perceived intensity rating. For example, a study from Coltell et al. did show an inverse association between higher taste intensity and MetS components [32].

As mentioned above, our subjects have reported adding salt to their food before tasting quite frequently, and as many as 32% of subjects said they do it often or almost always. Furthermore, it is known that the average daily salt intake in Croatian population is extremely high, estimated to be as much as 13 g in men and 10 g in women [73]. This could have affected our results substantially. Indeed, previous studies demonstrated plasticity in salt taste perception, pointing to the findings that manipulation with dietary salt intake has the potential to change both salt preference and perception in adults [73]. Once people are habituated to a diet with the certain amount of salt, foods with lower salt content are perceived as less intense. Within the context of a high salt diet, this may lead to the poor acceptance of low salt foods, which may explain in part why adherence to a low sodium diet is initially difficult for most of the people [74]. Additionally, salt taste perception can be influenced by many other characteristics, such as smoking, excessive alcohol intake and age [75].

Our study showed that subjects with higher salt taste threshold were on average older than those with lower threshold. Age was also negatively correlated with salt taste intensity perception. The same was found in a recent study including a large sample from the general population, with even stronger negative association for the higher concentrations of the testants [76]. It is well established that sensory acuity diminishes with age, albeit the sense of smell is more prone to deterioration with age than the sense of taste [77]. Consequently, age-related changes in sensory perception and preference could have a major impact on appetite and food intake. In addition, ageing brings a variety of lifestyle changes as well as a greater number of chronic diseases and associated medications, which can affect taste sensation [78].

Several limitations and advantages of this study should be considered. Firstly, this is a cross-sectional study, and therefore causality cannot be assessed in determining taste differences as a cause of MetS. Namely, we cannot elucidate whether subjects with reduced salt taste acuity (increased threshold) developed MetS as a consequence of this sensory characteristic or the MetS was a causal predictor of the salt taste acuity loss. For example, subjects with lower salt taste acuity could also have altered perception of food, which could influence their food choices toward less healthy, more processed salty foods, influencing their body weight and the risk for MetS development. Unfortunately, these habits were not included in the questionnaire, and we can not clarify this further based on the available data. Secondly, data for salt taste intensity and hedonic perception were available only for a subset of subjects, which might have introduced bias of undetermined direction and magnitude. The conditions of the field-testing were not ideal as would have been in the laboratory, which could have resulted in less accurate and precise measures of salt taste perception. Lastly, we used aqueous solution to deliver testant, while most of the salty foods are not consumed in a liquid form, which might have influenced suprathreshold responses and especially hedonic responses of the subjects. For instance, subjects could have rated concentrated salty solution as more intense and less hedonically appealing, compared to the rating they would provide for solid salty foods, such as chips or cured processed meats. This could be behind our result of absent association between suprathreshold salt taste perception and prevalent MetS and its components. The main advantages of this study include a large population-based sample size and the testing of both threshold and suprathreshold salt taste perception. The analysis included numerous confounders related to MetS, such as diet, physical activity, smoking, alcohol consumption and adding salt before tasting food. To the best of our knowledge, this is the first study to examine the association between salt taste threshold sensitivity, intensity perception and hedonic rating, with both MetS and the Mediterranean diet.

In conclusion, this study adds new insights into the existing body of knowledge about salt taste perception, nutrition and possible health outcomes. Still, there is much to be investigated, given the amount of discrepancies between previous studies, which are limited in number. Given the importance of salt taste in food palatability and associated food choices, as well as its role in propelling overweight and obesity, hypertension, MetS and other health outcomes, further studies are warranted.

Author Contributions: I.K., C.H., and O.P. conceived and designed this study; C.H. and O.P. secured the funding; A.P., I.P., C.H., O.P. and I.K. performed data acquisition; A.P. and I.K. analyzed the data; N.N.V., L.M., R.D., T.M. and I.K. interpreted the results and wrote the draft of the paper; all authors participated in discussions leading to the intellectual content and approved the final version for publication. All authors agree to be accountable for all aspects of the work in ensuring that questions related to the accuracy or integrity of any part of the work are appropriately investigated and resolved. All authors have read and agreed to the published version of the manuscript.

Funding: Croatian National Centre of Research Excellence in Personalized Healthcare grant (KK.01.1.1.01.0010); Centre of Competence in Molecular Diagnostics (KK.01.2.2.03.0006).

Acknowledgments: The data used in this study were partly used for the graduating thesis of the co-author Lana Mucalo, with substantial differences in sample size and data analysis compared to this study. Thesis reference: Mucalo, L. Salty taste perception and metabolic syndrome in the population of Dalmatia. Graduating Thesis, University of Split School of Medicine, Split, 2019-10-28. We would like to acknowledge the staff of Institute for Anthropological Research from Zagreb, Croatia for the support in the fieldwork.

Conflicts of Interest: The authors declare no conflict of interest.

Appendix A

Table A1. Subjects' characteristics according to the presence or absence of the metabolic syndrome.

	Metabolic Syndrome Absent $n = 1543$	Metabolic Syndrome Present $n = 1254$	p
	Socio-demographic characteristics		
Age (years); median (IQR)	49.0 (23.0)	61.0 (16.0)	<0.001 *
	Gender; n (%)		
Women	1000 (64.8)	756 (60.3)	0.014 †
Men	543 (35.2)	498 (39.7)	
	Place of residence; n (%)		
Vis	149 (9.7)	241 (19.2)	<0.001 †
Korčula	1095 (71.0)	812 (64.8)	
Split	299 (19.4)	201 (16.0)	

Table A1. *Cont.*

	Metabolic Syndrome Absent *n* = 1543	Metabolic Syndrome Present *n* = 1254	*p*
	Education level; *n* (%)		
Primary school	225 (14.8)	424 (34.1)	<0.001 †
Secondary school	874 (57.3)	557 (44.8)	
University level	426 (27.9)	263 (21.1)	
	Anthropometry		
Weight (kg); median (IQR)	73.0 (19.4)	82.4 (21.3)	<0.001 *
BMI (kg/m^2); median (IQR)	24.4 (5.1)	27.8 (5.3)	<0.001 *
WHR; median (IQR)	0.88 (0.10)	0.95 (0.09)	<0.001 *
WHtR; median (IQR)	0.52 (0.08)	0.59 (0.08)	<0.001 *
	Metabolic syndrome components		
Waist circumference (mm); median (IQR)	890.0 (150.0)	990.0 (129.5)	<0.001 *
Fasting glucose (mmol/L); median (IQR)	5.0 (0.7)	5.8 (1.2)	<0.001 *
Systolic blood pressure; median (IQR)	120.0 (15.0)	140.0 (20.0)	<0.001 *
Diastolic blood pressure; median (IQR)	80.0 (10.0)	81.0 (10.0)	<0.001 *
Triglycerides (mmol/L); median (IQR)	1.0 (0.5)	1.6 (1.0)	<0.001 *
HDL (mmol/L); median (IQR)	1.5 (0.4)	1.3 (0.4)	<0.001 *
	Other biochemical parameters		
Total cholesterol (mmol/L); median (IQR)	5.6 (1.6)	6.1 (1.6)	<0.001 *
LDL cholesterol (mmol/L); median (IQR)	3.6 (1.4)	3.9 (1.4)	<0.001 *
HbA1c (%); median (IQR)	5.2 (0.5)	5.5 (0.7)	<0.001 *
	Habits		
Smoking; *n* (%)			
Never-smokers	719 (47.0)	626 (50.4)	<0.001 †
Ex-smokers	326 (21.3)	327 (26.3)	
Active smokers	485 (31.7)	290 (23.3)	
Pack-years in smokers; median (IQR)	10.5 (17.5)	22.1 (25.3)	<0.001 *
Alcohol intake (units/week); median (IQR)	6.8 (18.9)	6.7 (21.5)	0.086 *
Physical activity; *n* (%)			
Low	328 (21.5)	300 (24.4)	0.004 †
Moderate	1066 (69.9)	788 (64.2)	
Intensive	130 (8.5)	139 (11.3)	
Mediterranean diet adherence (MDSS points); median (IQR)	11.0 (5.0)	12.0 (5.0)	<0.001 *
	Adding salt before tasting food; *n* (%)		
Never	635 (43.4)	438 (37.9)	0.012 †
Occasionally	373 (25.5)	295 (25.5)	
Often	350 (23.9)	320 (27.7)	
Almost always	105 (7.2)	104 (9.0)	

IQR—interquartile range; BMI—body mass index; WHR—waist-to-hip ratio; WHtR—waist-to-height ratio; MDSS—Mediterranean Diet Serving Score; * Mann-Whitney U test; † chi-square test.

Table A2. Salt taste threshold and suprathreshold perception according to the presence or absence of the metabolic syndrome, stratified by age groups.

	18–34.9 Years Old			35–64.9 Years Old			≥65 Years Old		
	Metabolic Syndrome Absent	**Metabolic Syndrome Present**	p †	**Metabolic Syndrome Absent**	**Metabolic Syndrome Present**	p †	**Metabolic Syndrome Absent**	**Metabolic Syndrome Present**	p †
Salt taste threshold (sample size is 2797)	$n = 342$	$n = 48$		$n = 980$	$n = 731$		$n = 221$	$n = 475$	
Lower salt taste threshold/higher acuity; *n* (%)	182 (53.2)	29 (50.4)		423 (43.2)	282 (38.6)		69 (31.2)	108 (22.7)	
Intermediate salt taste threshold/acuity; *n* (%)	134 (39.2)	15 (31.3)	0.570	439 (44.8)	313 (42.8)	0.001	99 (44.8)	236 (49.7)	0.056
Higher salt taste threshold/lower acuity; *n* (%)	26 (7.6)	4 (3.3)		118 (12.0)	136 (18.6)		53 (24.0)	131 (27.6)	
Salt taste suprathreshold (sample size is 1155)	$n = 171$	$n = 17$		$n = 440$	$n = 257$		$n = 97$	$n = 173$	
				Salt taste intensity perception					
No sensation to medium strong; *n* (%)	13 (7.6)	0 (0.0)		40 (9.1)	18 (7.0)		11 (11.3)	21 (12.1)	
Strong to very strong; N (%)	90 (52.6)	11 (64.7)	0.408	293 (66.6)	177 (68.9)	0.614	66 (68.0)	129 (74.6)	0.287
Extremely strong; *n* (%)	68 (39.8)	6 (35.3)		107 (24.3)	62 (24.1)		20 (20.6)	23 (13.3)	
				Salt taste hedonic perception					
Dislike; *n* (%)	129 (75.4)	13 (76.5)		342 (77.7)	183 (71.2)		79 (81.4)	126 (72.8)	
Neither like nor dislike; *n* (%)	25 (14.6)	3 (17.6)	0.834	70 (15.9)	54 (21.0)	0.152	14 (14.4)	28 (16.2)	0.125
Like; *n* (%)	17 (9.9)	1 (5.9)		28 (6.4)	20 (7.8)		4 (4.1)	19 (11.0)	

† chi-square test.

Table A3. Mediterranean diet and food groups consumption based on the modern Mediterranean pyramid [35] according to the salt taste recognition threshold perception, data are presented as n (%).

	Lower Salt Taste Threshold/Higher Acuity n = 1094	Intermediate Salt Taste Threshold/Acuity n = 1236	Higher Salt Taste Threshold/Lower Acuity n = 468	p †
Fruit (several times a day)	614 (56.1)	647 (52.3)	229 (48.9)	0.023
Vegetables (several times a day)	388 (35.5)	387 (31.3)	169 (36.1)	0.052
Olive oil intake (several times a day)	692 (63.3)	831 (67.2)	331 (70.7)	0.010
Cereals (several times a day)	984 (89.9)	1111 (89.9)	424 (90.6)	0.902
Nuts (1–2 times a day)	54 (4.9)	59 (4.8)	14 (3.0)	0.208
Dairy products (1–2 times a day)	256 (23.4)	255 (20.6)	100 (21.4)	0.262
Fish (≥2 times a week)	727 (66.5)	808 (65.4)	355 (75.9)	<0.001
Legumes (≥2 times a week)	264 (24.1)	270 (21.8)	133 (28.4)	0.017
Potatoes (≤3 times a week)	813 (74.3)	820 (66.3)	281 (60.0)	<0.001
Eggs (2–4 times a week)	278 (25.4)	294 (23.8)	111 (23.7)	0.614
White meat (2 times a week)	440 (40.2)	490 (39.6)	224 (47.9)	0.006
Red meat (<2 times a week)	318 (29.1)	332 (26.9)	129 (27.6)	0.490
Sweets (≤2 times a week)	287 (26.2)	364 (29.4)	130 (27.8)	0.224
Wine (1–2 glasses a day)	200 (18.3)	188 (15.2)	93 (19.9)	0.035
Overall Mediterranean diet compliance (MDSS ≥ 14 points)	275 (25.8)	258 (21.8)	114 (25.3)	0.062

† chi-square test.

References

1. Grundy, S.M. Metabolic syndrome pandemic. *Arter. Thromb. Vasc. Biol.* **2008**, *28*, 629–636. [CrossRef] [PubMed]
2. O'Neill, S.; O'Driscoll, L. Metabolic syndrome: A closer look at the growing epidemic and its associated pathologies. *Obes. Rev.* **2015**, *16*, 1–12. [CrossRef] [PubMed]
3. Aguilar, M.; Bhuket, T.; Torres, S.; Liu, B.; Wong, R.J. Prevalence of the metabolic syndrome in the United States, 2003–2012. *JAMA* **2015**, *313*, 1973–1974. [CrossRef] [PubMed]
4. Lan, Y.; Mai, Z.; Zhou, S.; Liu, Y.; Li, S.; Zhao, Z.; Duan, X.; Cai, C.; Deng, T.; Zhu, W.; et al. Prevalence of metabolic syndrome in China: An up-dated cross-sectional study. *PLoS ONE* **2018**, *13*. [CrossRef] [PubMed]
5. Raposo, L.; Severo, M.; Barros, H.; Santos, A.C. The prevalence of the metabolic syndrome in Portugal: The PORMETS study. *BMC Public Health* **2017**, *17*, 555. [CrossRef] [PubMed]
6. Iwasaki, Y.; Arisawa, K.; Katsuura-Kamano, S.; Uemura, H.; Tsukamoto, M.; Kadomatsu, Y.; Okada, R.; Hishida, A.; Tanaka, K.; Hara, M.; et al. Associations of Nutrient Patterns with the Prevalence of Metabolic Syndrome: Results from the Baseline Data of the Japan Multi-Institutional Collaborative Cohort Study. *Nutrients* **2019**, *11*, 990. [CrossRef] [PubMed]
7. Todowede, O.O.; Mianda, S.Z.; Sartorius, B. Prevalence of metabolic syndrome among HIV-positive and HIV-negative populations in sub-Saharan Africa-a systematic review and meta-analysis. *Syst. Rev.* **2019**, *8*, 4. [CrossRef]
8. Tocci, G.; Ferrucci, A.; Bruno, G.; Mannarino, E.; Nati, G.; Trimarco, B.; Volpe, M. Prevalence of metabolic syndrome in the clinical practice of general medicine in Italy. *Cardiovasc. Diagn. Ther.* **2015**, *5*, 271–279. [CrossRef]
9. Athyros, V.G.; Bouloukos, V.I.; Pehlivanidis, A.N.; Papageorgiou, A.A.; Dionysopoulou, S.G.; Symeonidis, A.N.; Petridis, D.I.; Kapousouzi, M.I.; Satsoglou, E.A.; Mikhailidis, D.P.; et al. The prevalence of the metabolic syndrome in Greece: The MetS-Greece Multicentre Study. *Diabetes Obes. Metab.* **2005**, *7*, 397–405. [CrossRef]
10. Fernandez-Berges, D.; Cabrera de Leon, A.; Sanz, H.; Elosua, R.; Guembe, M.J.; Alzamora, M.; Vega-Alonso, T.; Felix-Redondo, F.J.; Ortiz-Marron, H.; Rigo, F.; et al. Metabolic syndrome in Spain: Prevalence and coronary risk associated with harmonized definition and WHO proposal. DARIOS study. *Rev. Esp. Cardiol. (Engl. Ed.)* **2012**, *65*, 241–248. [CrossRef]

11. Relja, A.; Miljkovic, A.; Gelemanovic, A.; Boskovic, M.; Hayward, C.; Polasek, O.; Kolcic, I. Nut Consumption and Cardiovascular Risk Factors: A Cross-Sectional Study in a Mediterranean Population. *Nutrients* **2017**, *9*, 1296. [CrossRef] [PubMed]
12. Tucak-Zoric, S.; Curcic, I.B.; Mihalj, H.; Dumancic, I.; Zelic, Z.; Cetina, N.M.; Smolic, R.; Volarevic, M.; Missoni, S.; Tomljenovic, A.; et al. Prevalence of metabolic syndrome in the interior of Croatia: The Baranja region. *Coll. Antropol.* **2008**, *32*, 659–665. [PubMed]
13. Saklayen, M.G. The Global Epidemic of the Metabolic Syndrome. *Curr. Hypertens. Rep.* **2018**, *20*, 12. [CrossRef] [PubMed]
14. Afshin, A.; Forouzanfar, M.H.; Reitsma, M.B.; Sur, P.; Estep, K.; Lee, A.; Marczak, L.; Mokdad, A.H.; Moradi-Lakeh, M.; Naghavi, M.; et al. Health Effects of Overweight and Obesity in 195 Countries over 25 Years. *N. Engl. J. Med.* **2017**, *377*, 13–27. [CrossRef] [PubMed]
15. Garrido-Miguel, M.; Cavero-Redondo, I.; Alvarez-Bueno, C.; Rodriguez-Artalejo, F.; Moreno, L.A.; Ruiz, J.R.; Ahrens, W.; Martinez-Vizcaino, V. Prevalence and Trends of Overweight and Obesity in European Children From 1999 to 2016: A Systematic Review and Meta-analysis. *JAMA Pediatrics* **2019**. [CrossRef] [PubMed]
16. Leon-Munoz, L.M.; Guallar-Castillon, P.; Graciani, A.; Lopez-Garcia, E.; Mesas, A.E.; Aguilera, M.T.; Banegas, J.R.; Rodriguez-Artalejo, F. Adherence to the Mediterranean diet pattern has declined in Spanish adults. *J. Nutr.* **2012**, *142*, 1843–1850. [CrossRef]
17. Kolcic, I.; Relja, A.; Gelemanovic, A.; Miljkovic, A.; Boban, K.; Hayward, C.; Rudan, I.; Polasek, O. Mediterranean diet in the southern Croatia - does it still exist? *Croat. Med. J.* **2016**, *57*, 415–424. [CrossRef]
18. Santulli, G.; Pascale, V.; Finelli, R.; Visco, V.; Giannotti, R.; Massari, A.; Morisco, C.; Ciccarelli, M.; Illario, M.; Iaccarino, G.; et al. We Are What We Eat: Impact Of Food From Short Supply Chain On Metabolic Syndrome. *J. Clin. Med.* **2019**, *8*, 2061. [CrossRef]
19. Kargin, D.; Tomaino, L.; Serra-Majem, L. Experimental Outcomes of the Mediterranean Diet: Lessons Learned from the Predimed Randomized Controlled Trial. *Nutrients* **2019**, *11*, 2991. [CrossRef]
20. Esposito, K.; Maiorino, M.I.; Bellastella, G.; Chiodini, P.; Panagiotakos, D.; Giugliano, D. A journey into a Mediterranean diet and type 2 diabetes: A systematic review with meta-analyses. *BMJ Open* **2015**, *5*. [CrossRef]
21. Cox, D.N.; Perry, L.; Moore, P.B.; Vallis, L.; Mela, D.J. Sensory and hedonic associations with macronutrient and energy intakes of lean and obese consumers. *Int. J. Obes. Relat. Metab. Disord.* **1999**, *23*, 403–410. [CrossRef] [PubMed]
22. Donaldson, L.F.; Bennett, L.; Baic, S.; Melichar, J.K. Taste and weight: Is there a link? *Am. J. Clin. Nutr.* **2009**, *90*, 800S–803S. [CrossRef] [PubMed]
23. Li, Q.; Jin, R.; Yu, H.; Lang, H.; Cui, Y.; Xiong, S.; Sun, F.; He, C.; Liu, D.; Jia, H.; et al. Enhancement of Neural Salty Preference in Obesity. *Cell Physiol. Biochem.* **2017**, *43*, 1987–2000. [CrossRef] [PubMed]
24. Fernandez-Garcia, J.C.; Alcaide, J.; Santiago-Fernandez, C.; Roca-Rodriguez, M.M.; Aguera, Z.; Banos, R.; Botella, C.; de la Torre, R.; Fernandez-Real, J.M.; Fruhbeck, G.; et al. An increase in visceral fat is associated with a decrease in the taste and olfactory capacity. *PLoS ONE* **2017**, *12*. [CrossRef]
25. Pasquet, P.; Frelut, M.L.; Simmen, B.; Hladik, C.M.; Monneuse, M.O. Taste perception in massively obese and in non-obese adolescents. *Int. J. Pediatric Obes.* **2007**, *2*, 242–248. [CrossRef] [PubMed]
26. Kirsten, V.R.; Wagner, M.B. Salt taste sensitivity thresholds in adolescents: Are there any relationships with body composition and blood pressure levels? *Appetite* **2014**, *81*, 89–92. [CrossRef]
27. Lampure, A.; Castetbon, K.; Deglaire, A.; Schlich, P.; Peneau, S.; Hercberg, S.; Mejean, C. Associations between liking for fat, sweet or salt and obesity risk in French adults: A prospective cohort study. *Int. J. Behav. Nutr. Phys. Act.* **2016**, *13*, 74. [CrossRef]
28. Kaufman, A.; Choo, E.; Koh, A.; Dando, R. Inflammation arising from obesity reduces taste bud abundance and inhibits renewal. *PLoS Biol.* **2018**, *16*, e2001959. [CrossRef]
29. He, F.J.; Marrero, N.M.; MacGregor, G.A. Salt intake is related to soft drink consumption in children and adolescents:a link to obesity? *Hypertension* **2008**, *51*, 629–634. [CrossRef]
30. Grillo, A.; Salvi, L.; Coruzzi, P.; Salvi, P.; Parati, G. Sodium Intake and Hypertension. *Nutrients* **2019**, *11*, 1970. [CrossRef]
31. Bertoli, S.; Laureati, M.; Battezzati, A.; Bergamaschi, V.; Cereda, E.; Spadafranca, A.; Vignati, L.; Pagliarini, E. Taste sensitivity, nutritional status and metabolic syndrome: Implication in weight loss dietary interventions. *World J. Diabetes* **2014**, *5*, 717–723. [CrossRef] [PubMed]

32. Coltell, O.; Sorli, J.V.; Asensio, E.M.; Fernandez-Carrion, R.; Barragan, R.; Ortega-Azorin, C.; Estruch, R.; Gonzalez, J.I.; Salas-Salvado, J.; Lamon-Fava, S.; et al. Association between taste perception and adiposity in overweight or obese older subjects with metabolic syndrome and identification of novel taste-related genes. *Am. J. Clin. Nutr.* **2019**, *109*, 1709–1723. [CrossRef]
33. Mottillo, S.; Filion, K.B.; Genest, J.; Joseph, L.; Pilote, L.; Poirier, P.; Rinfret, S.; Schiffrin, E.L.; Eisenberg, M.J. The metabolic syndrome and cardiovascular risk a systematic review and meta-analysis. *J. Am. Coll. Cardiol.* **2010**, *56*, 1113–1132. [CrossRef] [PubMed]
34. Rudan, I.; Marusic, A.; Jankovic, S.; Rotim, K.; Boban, M.; Lauc, G.; Grkovic, I.; Dogas, Z.; Zemunik, T.; Vatavuk, Z.; et al. "10001 Dalmatians:" Croatia launches its national biobank. *Croat. Med. J.* **2009**, *50*, 4–6. [CrossRef] [PubMed]
35. Monteagudo, C.; Mariscal-Arcas, M.; Rivas, A.; Lorenzo-Tovar, M.L.; Tur, J.A.; Olea-Serrano, F. Proposal of a Mediterranean Diet Serving Score. *PLoS ONE* **2015**, *10*. [CrossRef]
36. Alberti, K.G.; Eckel, R.H.; Grundy, S.M.; Zimmet, P.Z.; Cleeman, J.I.; Donato, K.A.; Fruchart, J.C.; James, W.P.; Loria, C.M.; Smith, S.C., Jr. Harmonizing the metabolic syndrome: A joint interim statement of the International Diabetes Federation Task Force on Epidemiology and Prevention; National Heart, Lung, and Blood Institute; American Heart Association; World Heart Federation; International Atherosclerosis Society; and International Association for the Study of Obesity. *Circulation* **2009**, *120*, 1640–1645. [CrossRef]
37. Fernandez, M.D.M.; Bourquin, M.; Perou, M.; Programme, W.; Area, W.; Applications, I.; IRAM, I.; Fernandez, M. ISO/TC 34/SC 12 - Sensory analysis. Available online: https://www.iso.org/committee/47942 (accessed on 25 March 2020).
38. Webb, J.; Bolhuis, D.P.; Cicerale, S.; Hayes, J.E.; Keast, R. The Relationships Between Common Measurements of Taste Function. *Chemosens. Percept.* **2015**, *8*, 11–18. [CrossRef]
39. Green, B.G.; Dalton, P.; Cowart, B.; Shaffer, G.; Rankin, K.; Higgins, J. Evaluating the 'Labeled Magnitude Scale' for measuring sensations of taste and smell. *Chem. Senses* **1996**, *21*, 323–334. [CrossRef]
40. Schutz, H.G.; Cardello, A.V. A Labeled Affective Magnitude (Lam) Scale For Assessing Food Liking/Disliking. *J. Sens. Stud.* **2001**, *16*, 117–159. [CrossRef]
41. Ford, E.S.; Giles, W.H.; Dietz, W.H. Prevalence of the metabolic syndrome among US adults: Findings from the third National Health and Nutrition Examination Survey. *JAMA* **2002**, *287*, 356–359. [CrossRef]
42. Kolcic, I.; Vorko-Jovic, A.; Salzer, B.; Smoljanovic, M.; Kern, J.; Vuletic, S. Metabolic syndrome in a metapopulation of Croatian island isolates. *Croat. Med. J.* **2006**, *47*, 585–592. [PubMed]
43. Pecin, I.; Ivkovic, V.; Vrdoljak, A.; Karanovic, S.; Fucek, M.; Dika, Z.; Laganovic, M.; Premuzic, V.; Kos, J.; Zeljkovic-Vrkic, T.; et al. High prevalence of metabolic syndrome in rural part of croatia. Croatian rural study. *J. Hypertens.* **2015**, *33*. [CrossRef]
44. Simunovic, M.; Bozic, J.; Milic, L.; Unic, I.; Skrabic, V. The Prevalence of Metabolic Syndrome and Cardiovascular Risk Factors in Obese Children and Adolescents in Dalmatia: A Hospital Based Study. *Int. J. Endocrinol.* **2016**, *2016*, 1823561. [CrossRef] [PubMed]
45. Bergman Markovic, B.; Vrdoljak, D.; Kranjcevic, K.; Vucak, J.; Kern, J.; Bielen, I.; Ivezic Lalic, D.; Katic, M.; Reiner, Z. Continental-Mediterranean and rural-urban differences in cardiovascular risk factors in Croatian population. *Croat. Med. J.* **2011**, *52*, 566–575. [CrossRef] [PubMed]
46. Lakka, H.M.; Laaksonen, D.E.; Lakka, T.A.; Niskanen, L.K.; Kumpusalo, E.; Tuomilehto, J.; Salonen, J.T. The metabolic syndrome and total and cardiovascular disease mortality in middle-aged men. *JAMA* **2002**, *288*, 2709–2716. [CrossRef] [PubMed]
47. Hu, G.; Qiao, Q.; Tuomilehto, J.; Balkau, B.; Borch-Johnsen, K.; Pyorala, K.; Group, D.S. Prevalence of the metabolic syndrome and its relation to all-cause and cardiovascular mortality in nondiabetic European men and women. *Arch. Intern. Med.* **2004**, *164*, 1066–1076. [CrossRef] [PubMed]
48. Liem, D.G.; Russell, C.G. The Influence of Taste Liking on the Consumption of Nutrient Rich and Nutrient Poor Foods. *Front. Nutr.* **2019**, *6*, 174. [CrossRef]
49. Tan, S.Y.; Tucker, R.M. Sweet Taste as a Predictor of Dietary Intake: A Systematic Review. *Nutrients* **2019**, *11*, 94. [CrossRef]
50. Tepper, B.J.; Barbarossa, I.T. Taste, Nutrition, and Health. *Nutrients* **2020**, *12*, 155. [CrossRef] [PubMed]
51. Rust, P.; Ekmekcioglu, C. Impact of Salt Intake on the Pathogenesis and Treatment of Hypertension. *Adv. Exp. Med. Biol.* **2017**, *956*, 61–84. [CrossRef] [PubMed]

52. Pilic, L.; Mavrommatis, Y. Genetic predisposition to salt-sensitive normotension and its effects on salt taste perception and intake. *Br. J. Nutr.* **2018**, *120*, 721–731. [CrossRef] [PubMed]
53. Oh, S.W.; Han, K.H.; Han, S.Y.; Koo, H.S.; Kim, S.; Chin, H.J. Association of Sodium Excretion With Metabolic Syndrome, Insulin Resistance, and Body Fat. *Medicine (Baltimore)* **2015**, *94*. [CrossRef] [PubMed]
54. Skrandies, W.; Zschieschang, R. Olfactory and gustatory functions and its relation to body weight. *Physiol. Behav.* **2015**, *142*, 1–4. [CrossRef] [PubMed]
55. Hardikar, S.; Hochenberger, R.; Villringer, A.; Ohla, K. Higher sensitivity to sweet and salty taste in obese compared to lean individuals. *Appetite* **2017**, *111*, 158–165. [CrossRef]
56. Martinelli, J.; Conde, S.R.; Araujo, A.R.; Marcadenti, A. Association between salt taste sensitivity threshold and blood pressure in healthy individuals: A cross-sectional study. *Sao Paulo Med. J.* **2020**. [CrossRef]
57. Kubota, Y.; Higashiyama, A.; Sugiyama, D.; Nishida, Y.; Kubo, S.; Hirata, T.; Kadota, A.; Miyamatsu, N.; Wakabayashi, I.; Miyamoto, Y.; et al. Association between impairment of salty taste recognition and masked hypertension based on home blood pressure in Japanese residents: The KOBE study. *Hypertens. Res.* **2018**, *41*, 756–762. [CrossRef]
58. Rabin, M.; Poli de Figueiredo, C.E.; Wagner, M.B.; Antonello, I.C.F. Salt taste sensitivity threshold and exercise-induced hypertension. *Appetite* **2009**, *52*, 609–613. [CrossRef]
59. Kim, C.Y.; Ye, M.K.; Lee, Y.S. The salt-taste threshold in untreated hypertensive patients. *Clin. Hypertens.* **2017**, *23*, 22. [CrossRef]
60. Fischer, M.E.; Cruickshanks, K.J.; Pinto, A.; Schubert, C.R.; Klein, B.E.; Klein, R.; Nieto, F.J.; Pankow, J.S.; Snyder, D.J.; Keating, B.J. Intensity of Salt Taste and Prevalence of Hypertension Are Not Related in the Beaver Dam Offspring Study. *Chemosens. Percept.* **2012**, *5*, 139–145. [CrossRef]
61. Libuda, L.; Kersting, M.; Alexy, U. Consumption of dietary salt measured by urinary sodium excretion and its association with body weight status in healthy children and adolescents. *Public Health Nutr.* **2012**, *15*, 433–441. [CrossRef]
62. Donovan, D.S.; Solomon, C.G.; Seely, E.W.; Williams, G.H.; Simonson, D.C. Effect of sodium intake on insulin sensitivity. *Am. J. Physiol.* **1993**, *264 Pt 1*, E730–E734. [CrossRef]
63. Lanaspa, M.A.; Kuwabara, M.; Andres-Hernando, A.; Li, N.; Cicerchi, C.; Jensen, T.; Orlicky, D.J.; Roncal-Jimenez, C.A.; Ishimoto, T.; Nakagawa, T.; et al. High salt intake causes leptin resistance and obesity in mice by stimulating endogenous fructose production and metabolism. *Proc. Natl. Acad. Sci. USA* **2018**, *115*, 3138–3143. [CrossRef] [PubMed]
64. Kure Liu, C.; Joseph, P.V.; Feldman, D.E.; Kroll, D.S.; Burns, J.A.; Manza, P.; Volkow, N.D.; Wang, G.J. Brain Imaging of Taste Perception in Obesity: A Review. *Curr. Nutr. Rep.* **2019**, *8*, 108–119. [CrossRef] [PubMed]
65. Cattaneo, C.; Riso, P.; Laureati, M.; Gargari, G.; Pagliarini, E. Exploring Associations between Interindividual Differences in Taste Perception, Oral Microbiota Composition, and Reported Food Intake. *Nutrients* **2019**, *11*, 1167. [CrossRef] [PubMed]
66. Leong, C.S.; Forde, C.G.; Tey, S.L.; Henry, C.J. Taste perception and diet in people of Chinese ancestry. *Asia Pac. J. Clin. Nutr.* **2018**, *27*, 478–486. [CrossRef]
67. Dernini, S.; Berry, E.M.; Serra-Majem, L.; La Vecchia, C.; Capone, R.; Medina, F.X.; Aranceta-Bartrina, J.; Belahsen, R.; Burlingame, B.; Calabrese, G.; et al. Med Diet 4.0: The Mediterranean diet with four sustainable benefits. *Public Health Nutr.* **2017**, *20*, 1322–1330. [CrossRef]
68. Rehberg, J.; Stipcic, A.; Coric, T.; Kolcic, I.; Polasek, O. Mortality patterns in Southern Adriatic islands of Croatia: A registry-based study. *Croat. Med. J.* **2018**, *59*, 118–123. [CrossRef]
69. Estruch, R.; Ros, E.; Salas-Salvado, J.; Covas, M.I.; Corella, D.; Aros, F.; Gomez-Gracia, E.; Ruiz-Gutierrez, V.; Fiol, M.; Lapetra, J.; et al. Primary Prevention of Cardiovascular Disease with a Mediterranean Diet Supplemented with Extra-Virgin Olive Oil or Nuts. *N. Engl. J. Med.* **2018**, *378*. [CrossRef]
70. Franquesa, M.; Pujol-Busquets, G.; Garcia-Fernandez, E.; Rico, L.; Shamirian-Pulido, L.; Aguilar-Martinez, A.; Medina, F.X.; Serra-Majem, L.; Bach-Faig, A. Mediterranean Diet and Cardiodiabesity: A Systematic Review through Evidence-Based Answers to Key Clinical Questions. *Nutrients* **2019**, *11*, 655. [CrossRef]
71. Godos, J.; Zappala, G.; Bernardini, S.; Giambini, I.; Bes-Rastrollo, M.; Martinez-Gonzalez, M. Adherence to the Mediterranean diet is inversely associated with metabolic syndrome occurrence: A meta-analysis of observational studies. *Int. J. Food Sci. Nutr.* **2017**, *68*, 138–148. [CrossRef]

72. Castro-Barquero, S.; Tresserra-Rimbau, A.; Vitelli-Storelli, F.; Domenech, M.; Salas-Salvado, J.; Martin-Sanchez, V.; Rubin-Garcia, M.; Buil-Cosiales, P.; Corella, D.; Fito, M.; et al. Dietary Polyphenol Intake is Associated with HDL-Cholesterol and A Better Profile of other Components of the Metabolic Syndrome: A PREDIMED-Plus Sub-Study. *Nutrients* **2020**, *12*, 689. [CrossRef] [PubMed]
73. Jelakovic, B.; Kaic-Rak, A.; Milicic, D.; Premuzic, V.; Skupnjak, B.; Reiner, Z. [Less salt–more health. Croatian action on salt and health (CRASH)]. *Lijec. Vjesn.* **2009**, *131*, 87–92. [PubMed]
74. Bobowski, N. Shifting human salty taste preference: Potential opportunities and challenges in reducing dietary salt intake of Americans. *Chemosens. Percept.* **2015**, *8*, 112–116. [CrossRef] [PubMed]
75. Lampure, A.; Schlich, P.; Deglaire, A.; Castetbon, K.; Peneau, S.; Hercberg, S.; Mejean, C. Sociodemographic, psychological, and lifestyle characteristics are associated with a liking for salty and sweet tastes in French adults. *J. Nutr.* **2015**, *145*, 587–594. [CrossRef] [PubMed]
76. Barragan, R.; Coltell, O.; Portoles, O.; Asensio, E.M.; Sorli, J.V.; Ortega-Azorin, C.; Gonzalez, J.I.; Saiz, C.; Fernandez-Carrion, R.; Ordovas, J.M.; et al. Bitter, Sweet, Salty, Sour and Umami Taste Perception Decreases with Age: Sex-Specific Analysis, Modulation by Genetic Variants and Taste-Preference Associations in 18 to 80 Year-Old Subjects. *Nutrients* **2018**, *10*, 1539. [CrossRef]
77. Murphy, C. Nutrition and chemosensory perception in the elderly. *Crit. Rev. Food Sci. Nutr.* **1993**, *33*, 3–15. [CrossRef] [PubMed]
78. Schiffman, S.S. Influence of medications on taste and smell. *World J. Otorhinolaryngol. Head Neck Surg.* **2018**, *4*, 84–91. [CrossRef]

nutrients

MDPI

Article

Effects of Doenjang, a Traditional Korean Soybean Paste, with High-Salt Diet on Blood Pressure in Sprague–Dawley Rats

Eun-Gyung Mun, Jung Eun Park and Youn-Soo Cha *

Department of Food Science and Human Nutrition, Jeonbuk National University, 567 Baekje-daero, Deokjin-gu, Jeonju-si, Jeollabuk-do 54896, Korea; egmun1982@gmail.com (E.-G.M.); jpark@jbnu.ac.kr (J.E.P.)

* Correspondence: cha8@jbnu.ac.kr; Tel.: +82-63-270-3822

Received: 2 October 2019; Accepted: 7 November 2019; Published: 12 November 2019

check for updates

Abstract: Fermented foods in Korea contain a lot of salt. Although salt is reported to exacerbate health trouble, fermented foods have beneficial effects. We hypothesized that doenjang could reduce blood pressure in Sprague–Dawley (SD) rats fed a high-salt diet. Eighteen SD rats were divided into three groups: normal-salt (NS) group, high-salt (HS) group, and high-salt with doenjang (HSD) group. The salinity of doenjang and saltwater was adjusted to 8% using Mohr's method. Blood pressure was significantly reduced in the HSD group compared with the HS group. Water intake and urine excretion volume has significantly increased in the HS group compared with the HSD group. The excreted concentrations of urine sodium, urine potassium, and feces potassium significantly increased in the HSD group compared with the HS and NS groups. Renin level was significantly decreased in the HSD group compared to the other groups. These results indicate that eating traditional salty fermented food is not a direct cause of hypertension, and the intake of doenjang in normal healthy animals improved blood pressure.

Keywords: high-salt diet; blood pressure; doenjang; renin; soybean paste

1. Introduction

Doenjang, Korean soy paste, is a fermented soybean product in traditional Korean cuisine. Doenjang is the basis of various dishes such as soup, stew, and salad dressing. Doenjang is produced by fermentation with *Bacillus subtilis* and mold such as *Aspergillus*, *Rhozopus*, and *Mucor* species. Doenjang has a long fermentation period that extends from 2 to 24 months. Due to its fermentation, doenjang contains a high level of bioactive compounds such as isoflavones and saponins [1]. In addition, doenjang is a good source of essential amino acids, minerals, vitamins, and phenolic compounds [2,3].

In previous studies, doenjang showed anti-obesity effects in rat fed with high-fat diet [4], and overweight adults [5–7], anti-diabetic effects in high-fat diet-induced obese mice [8,9], anti-cancer effects in carcinogenic mice [10,11], and anti-inflammation effects in mice [12–14]. The intake of doenjang reduces body fat mass in overweight subjects [5] and decreases visceral or abdominal fat mass in overweight subjects with mutant alleles of PPAR-gamma or uncoupling protein-1 [6,7]. In the long term, the intake of fermented doenjang improved glucose tolerance and inhibited hyperglycemia in high-fat diet-induced obese mice [8,9]. These health benefits may be due to the higher content of aglycone isoflavones and the diversity and abundance of *Bacillus* probiotic strains [4,9].

Despite the health benefits, doenjang fermentation requires a large amount of salt. Doenjang is traditionally made from fermented meju that has been dipped in a 18%–20% salt solution for more than 30 days [15]. Doenjang is fermented with 4 kg of meju dipped in brine, 10 L of water, and 1.25 kg of salt; the final salinity of traditional doenjang is 12% [2].

The main dietary sources of sodium in Korea are kimchi, salt, soy sauce, and soybean paste [16]. The World Health Organization recommends adults eat less than 2000 mg of sodium or 5 g of salt every day. Elevated sodium levels can raise blood pressure and may increase the risk of heart disease and stroke [17]. Thus, reducing sodium intake is recommended in the guidelines of many organizations. The optimal level of sodium intake is controversial; however, the sodium reduction program will help people prevent cardiovascular disease in communities that eat large amounts of salt [18]. However, although there are few studies that have combined doenjang and salt, there is a lack of research on mechanisms to prevent the side effects of salt. We hypothesized that consumption of traditional fermented foods would offset the side effects of table salt intake. Therefore, the antihypertensive effects of doenjang, a traditional fermented soybean paste, on the regulation of blood pressure were investigated in Sprague–Dawley (SD) rats.

2. Materials and Methods

2.1. Preparation of Doenjang

Doenjang was supplied by SunChangJangLye Co., Ltd. (Sunchang-gun, Korea). Meju, soybean fermented with *Aspergillus oryzae* and *Bacillus subtilis* for one month, and saltwater (26%, *w*/*v*) were mixed at a 1:3 ratio and fermented for 2 months. When the meju fermentation is complete, it is separated into the liquid phase and the solid phase. The crushed solid phase was matured for 6 months (Figure 1). Matured doenjang was stored by freeze drying. Freeze-dried doenjang and salt (Samchun Chemical, Seoul, Korea) were dissolved in distilled water to a salinity of 8% [19–21] using Mohr's method [22].

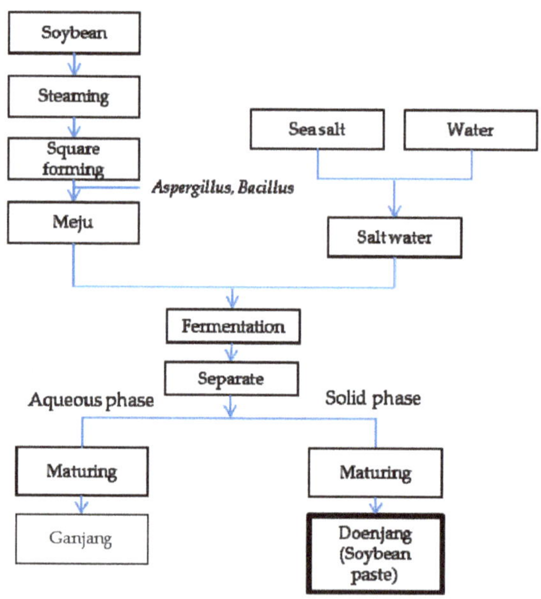

Figure 1. A manufacturing process diagram of doenjang. Meju, block of fermented soybeans; sea salt, manufacture of common salt by solar heat.

2.2. Animal Study

Male SD rats, aged three weeks, were purchased from Central Lab Animal Inc. (Seoul, Korea). The animals were fed on an AIN-76 diet (standard diet, Research Diets, New Brunswick, NY, USA) for one week, and eighteen rats were randomly divided into three groups ($n = 6$): normal-salt group

(0.3%, NS), high-salt group (8%, HS), and high-salt with doenjang group (HSD). Animals were kept at a temperature of 24 ± 2 °C with a humidity of 60 ± 5% and a light/dark cycle of 12:12 h. They were given free access to the AIN-76 diet and tap water.

The rats were fed orally administered 10 mL/kg body weight (BW) for the eight-week experimental period (Table 1). From the fifth to seventh weeks of the experimental period, rats were housed in metabolic cages for 24 h a week. Urine and feces were collected for analysis. The Animal and Use Committee of Chonbuk National University approved the experimental protocol (CBNU 2017-0016).

Table 1. Ingredients administered orally during the experimental period.

Group [1]	Energy (kcal)	Ash (g)	Crud Protein (g)	Carbohydrate (g)	Crud Fat (g)	Salt (%)	Sodium (mg/kg BW/day)	Potassium (mg/kg BW/day)
NS	0	0	0	0	0	0	0	0
HS	0	0	0	0	0	8.00	314.72	0
HSD	72.75	7.37	4.80	4.24	4.06	8.02	229.29	14.98

Body weight (BW). [1] The normal-salt (NS) group was fed with distilled water. The high-salt (HS) group was fed with saltwater that dissolved NaCl in distilled water. The high-salt with doenjang (HSD) group was fed with doenjang solution that mixed freeze-dried deonjang in distilled water.

2.3. Collection of Serum and Organs

After 12 h of overnight fasting, rats were anesthetized with 2 mg/kg BW of alfaxan (Jurox, Australia) and 0.5 mL/kg BW of rompun (Bayer, Seoul, Korea) by intramuscular injection, and blood was collected by orbital vein puncture. Serum was centrifuged at 1100× *g* for 15 min at 4 °C. Liver and kidney were harvested, rinsed, and weighed. Both tissues and serum were stored at −80 °C until analysis.

2.4. Measurement of Blood Pressure

Blood pressure was assessed weekly by the tail-cuff method (BP-2000; Visitech Systems, Inc., Apex, NC, USA) after 6 h of oral administration.

2.5. Serum Profile Analysis

Serum glutamate oxaloacetate transaminase (GOT) and glutamate pyruvate transaminase (GPT) were analyzed using a commercially available kit (Asan Pharmaceutical Co., Seoul, Korea).

Renin, angiotensin II (Ang II), and aldosterone in serum were measured by ELISA using a Rat Renin ELISA Kit (MyBioSource, San Diego, CA, USA), Angiotensin II ELISA Kit, and Aldosterone ELISA Kit (Enzo Life Sciences, Inc., Farmingdale, NY, USA).

2.6. Ion Content Analysis in Feces and Urine

Sodium and potassium ion concentrations in feces and urine on the 7th week were analyzed by inductively coupled plasma-optical emission spectrometry (ICP-OES; Optima 8300DV, Perkin Elmer, Waltham, MA, USA) at Wonkwang University Wonnature (Iksan, Korea).

2.7. Gene Expression in the Kidney Cortex

Total RNA was extracted from the kidney cortex using Trizol reagent (Invitrogen Life Technologies, Carlsbad, CA, USA), and the concentration was measured with BioDrop (Biochrom, Holliston, MA, USA). The RNA was reversed to complementary DNA (cDNA) using a high-capacity cDNA reverse transcription kit (Applied Biosystems, Foster City, CA, USA). The expression levels were quantified with real-time PCR using SYBR Green PCR Master Mix (Toyobo, Osaka, Japan) and a 7500 Real-Time PCR system (Applied Biosystems). The relative gene expression levels were analyzed by the 2-ΔΔCt (-delta delta comparative threshold) method using β-actin as the reference gene. Primer sequences for the angiotensin II type 1 receptor (AT1 receptor, sense 5′-ACTCTTTCCTACCGCCCTTC-3′, antisense 5′-TTAGCCCAAATGGTCCTCTG-3′), angiotensin-converting enzyme (ACE, sense 5′-GAGCCATCCTTCCCTTTTTC-3′, antisense 5′-GGCTGCAGCTCCTGGTATAG-3′), mineralocorticoid receptor (MR, sense 5′-GCTTTGATGGTAGCTGCG-3′, antisense 5′-TGAGCACCAATCCGGTAG-3′), Na^+/K^+ ATPase alpha 1 (NKAα1, sense 5′-CCGGAATTCTGCCTTCCCCTACTCCCTTCTCATC-3′, antisense 5′-TGCTCTAGACTTCCCCGCTGTCGTCCCCGTCCAC-3′), Na^+/H^+ exchanger3 (NHE3, sense 5′-GGAACAGAGGCGGAGGAGCAT-3′, antisense 5′-GAAGTTGTGTGCCAGATTCT-3′), Na^+/Ca^{2+} exchanger (NCX, sense 5′-GCGATTGCTTGTCTCGGGTC-3′, antisense 5′-CCACAGGTGTCCTCAAAGTCC-3′), Na^+/HCO_3^- co-exchanger (NBC, sense 5′-GGCACAGAGAGAGGAGGCTT-3′, antisense 5′-TGTCTTCCCAATGTCAGCCAG-3′) were used.

2.8. Analysis of Microbial Communities in Doenjang

Chunlab, Inc. (Seoul, Korea) performed the microbial community analysis of doenjang.

2.9. Statistical Analysis

The data were analyzed using one-way ANOVA with SPSS version 12.0 (SPSS Inc., Chicago, IL, USA). Values are expressed as mean ± standard deviation. The differences among groups were assessed using the Duncan's multiple range tests. Statistical significance was considered at $p < 0.05$.

3. Results

3.1. Metabolic Characterization and Serum Chemistry

The initial BW, final BW and feed intake were not significantly different in all groups. The water intake and urine volume were increased in the HS group compared with the HSD group. Although salt intake was the same, water intake was decreased in the HSD group (Table 2).

Table 2. Metabolic characterization in Sprague–Dawley (SD) rats fed with high-salt diet.

Group [1]	Initial Body Weight (g)	Final Body Weight (g)	Diet Intake (g/day)	Water Intake (mL/day)	Urine Volume (mL/day)
NS	128.24 ± 8.27	358.86 ± 33.18	16.97 ± 2.97	22.65 ± 4.81 [a,b]	9.54 ± 3.24 [c]
HS	129.40 ± 7.28	359.73 ± 26.51	19.35 ± 5.61	25.96 ± 6.4 [a]	16.17 ± 3.02 [a]
HSD	129.51 ± 9.03	356.09 ± 47.93	18.03 ± 3.19	18.88 ± 4.53 [b]	13.25 ± 2.53 [b]

Values are the mean ± standard deviation, with different letters significantly different ($p < 0.05$) by Duncan's multiple range test. Six rats were assigned to each group. [1] NS, normal-salt group; HS, high-salt group; HSD, high-salt with doenjang group.

Liver and kidney weights were not significantly different in all groups. In addition, liver-to-BW and kidney-to-BW ratios were no significant differences among groups. Significant difference was not observed in serum GOT and GPT levels among groups (Table 3).

Table 3. Organ weights and serum chemistry in SD rats fed with high-salt diet.

Group [1]	Liver Weight (g)	Kidney Weight (g)	Liver/BW	Kidney/BW	GOT (IU/L)	GPT (IU/L)
NS	11.55 ± 1.01	2.65 ± 0.34	3.23 ± 0.19 [b]	0.74 ± 0.06	25.33 ± 9	2.62 ± 2.35
HS	12.7 ± 1.13	2.81 ± 0.24	3.53 ± 0.18 [a]	0.78 ± 0.06	24.75 ± 4.81	2.31 ± 0.77
HSD	11.88 ± 2.06	2.7 ± 0.37	3.33 ± 0.26 [b]	0.76 ± 0.08	26.53 ± 4.88	2.31 ± 2.04

Values are the mean ± standard deviation, with different letters significantly different ($p < 0.05$) by Duncan's multiple range test. Six rats were assigned to each group. [1] NS, normal-salt group; HS, high-salt group; HSD, high-salt with doenjang group; BW, body weight; GOT, Glutamate oxaloacetate transaminase; GPT, glutamate pyruvate transaminase.

3.2. Systolic Blood Pressure

Initial systolic blood pressure was the same in all three groups. From the first to eighth week of the feeding period, the systolic blood pressure in the HS group was higher than in the other groups. However, significant differences were not observed in systolic blood pressure between the NS and HSD groups at the end of the experiment (Figure 2). Final systolic blood pressure was 157.45 ± 14.36 mmHg in HS, 135.44 ± 13.91 mmHg in HSD, and 137.14 ± 10.44 mmHg in NS groups.

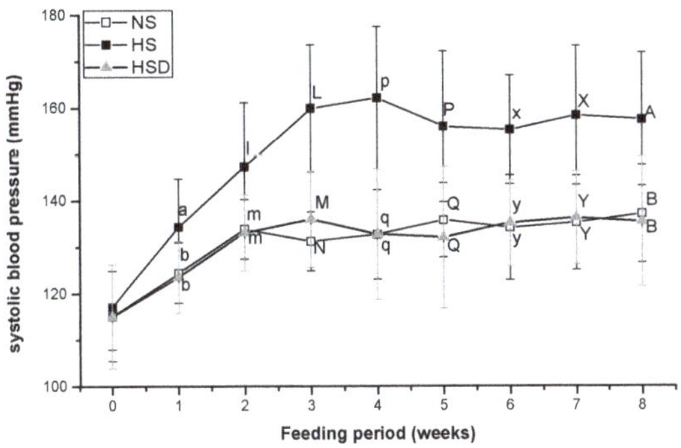

Figure 2. Changes in systolic blood pressure in Sprague–Dawley (SD) rats fed with high-salt diet. Values are the mean ± standard deviation, with different letters significantly different ($p < 0.05$) by Duncan's multiple range test. Six rats were assigned to each group. NS normal-salt group; HS, high-salt group; HSD, high-salt with doenjang group.

3.3. Ion Concentration in Feces and Urine

Sodium and potassium levels in feces and urine are shown in Figure 3. The sodium concentration in feces was not significant in three different groups. Fecal and urine potassium concentration was significantly higher in the HSD group than in the HS and NS groups. Sodium excretion in urine was significantly higher than NS in HS and HSD groups with a large amount of salt ingested.

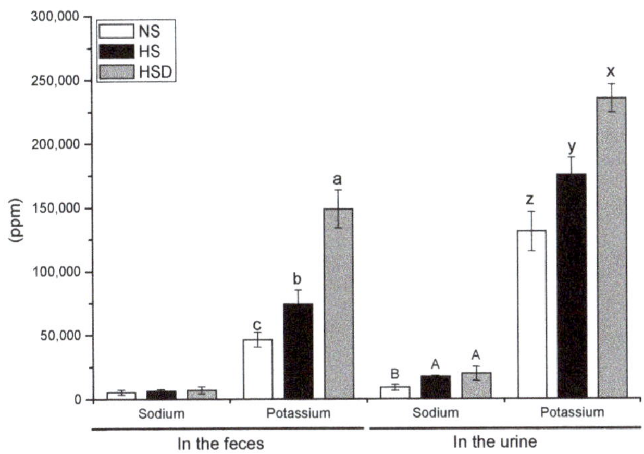

Figure 3. Ion concentrations in feces and urine in SD rats fed with high-salt diet. Values are the mean ± standard deviation, with different letters significantly different ($p < 0.05$) by Duncan's multiple range test. Six rats were assigned to each group. NS, normal-salt group; HS, high-salt group; HSD, high-salt with doenjang group.

3.4. Renin-Angiotensin-Aldosterone Levels in Serum

The serum renin-angiotensin-aldosterone (RAA) levels were analyzed (Table 4). Renin and aldosterone concentrations were slightly and significantly decreased in the HSD group compared to the HS group. Ang II level was significantly increased in the HS and HSD groups compared with the NS group.

Table 4. Renin-angiotensin-aldosterone (RAA) levels in serum (pg/mL).

Group [1]	Renin	Angiotensin II	Aldosterone
NS	41.72± 1.77 [a]	109.29 ± 5.19 [b]	19.88 ± 1.10 [a,b]
HS	43.10± 3.36 [a]	121.60 ± 4.98 [a]	21.37 ± 0.97 [a]
HSD	38.22± 2.66 [b]	118.85 ± 8.01 [a]	19.23 ± 2.03 [b]

Values are the mean ± standard deviation, with different letters significantly different ($p < 0.05$) by Duncan's multiple range test. Six rats were assigned to each group. [1] NS, normal-salt group; HS, high-salt group; HSD, high-salt with doenjang group; RAA, renin-angiotensin-aldosterone.

3.5. Relative Gene Expression in the Kidney Cortex

The mRNA expression of the kidney cortex was analyzed (Figure 4). The mRNA expression of Ang II type 1 (AT1) receptor, Ang-converting enzyme (ACE), mineralocorticoid receptor (MR), and Na^+/Ca^{2+} exchanger (NCX) was significantly higher in the HS group than in the HSD groups. Na^+/HCO_3^- co-exchanger (NBC) mRNA expression was significantly lower in the HS group than in the NS group. However, mRNA expression of the Na^+/K^+ ATPase α1 (NKAα1) and Na^+/H^+ exchanger 3 (NHE3) was not significantly in three different groups. Genetic changes in the HSD group were not significantly different from those in the NS group.

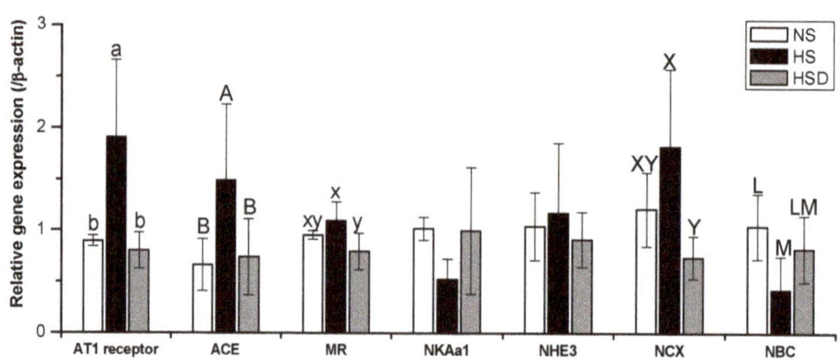

Figure 4. The mRNA expression of the kidney cortex in SD rats fed with high-salt diet. Values are the mean ± standard deviation, with different letters significantly different ($p < 0.05$) by Duncan's multiple range test. Six rats were assigned to each group. NS, normal-salt group; HS, high-salt group; HSD, high-salt with doenjang group; AT1 receptor, angiotensin II type 1 receptor; ACE, angiotensin-converting enzyme; MR, mineralocorticoid receptor; NKAα1, Na^+/K^+ ATPase alpha 1; NHE3, $Na^{+/}H^+$ exchanger3; NCX, Na^+/Ca^{2+} exchanger; NBC, $Na^{+/}HCO^{3-}$ co-exchanger.

3.6. Microbial Community in Doenjang

The microbial community in doenjang used in this experiment is shown in Figure 5 and consisted of the following: *Bacillales* (85.93%), *Lactobacillales* (13.58%), *Oceanospirillales* (0.31%), *Rhizobiales* (0.12%), *Enterobacteriales* (0.03%), *Clostridiales* (0.01%), *Desulfovibrionales* (0.01%), *Bacilli_uc* (0.01%), and *Bacteroidales* (0.01%). The most common species in doenjang were *Bacillus paralicheniformis* (69.38%), *Bacillus subtilis* group (5.18%), *Bacillus acidicola* (4.11%), *Bacillus_uc* (3.57%), and *Bacillus dabaoshanensis* (1.62%).

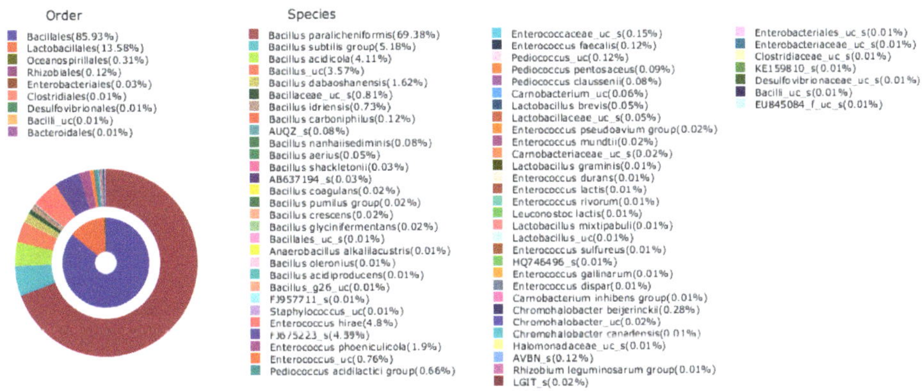

Figure 5. Microbial community in doenjang.

4. Discussion

Doenjang, a fermented soybean paste, is made from the fermentation of soybeans, salt, and water using a traditional method. The first step in making doenjang is to prepare meju by steeping, steaming, and forming the soybeans [10]. The second step is to soak fermented soybeans in saltwater and ferment them in natural conditions for one or two months [23]. The solid is separated from the liquid, which is collected to make the doenjang. The cooked cereals, salt, and crushed meju are added and left to ripen for 3–6 months [24]. Doenjang made from soybean and salt is traditionally used as both a condiment and has been used as a portion of health food that has anti-obesity, anti-diabetic, anti-cancer, and anti-inflammatory activities. Doenjang has ACE inhibitory effects that can help prevent increased blood pressure [24,25]. *Lactobacillus rhamnosus* in doenjang showed vigorous proteolytic activity and could aid in generating bioactive peptides [26].

In the present study, the intake of doenjang decreased blood pressure through modulation of the RAA system (RAAS), known to regulate blood pressure and fluid and electrolyte balance [27]. The RAAS plays a central role in regulating blood pressure by maintaining sodium and water homeostasis and vascular tone. Renin, as an inactive form, is synthesized from the kidney and released into the circulatory system in response to low levels of sodium in the tubular, low blood pressure in the arterioles of a renal glomerulus, and sympathetic activation. The active renin facilitates the division of angiotensinogen, which is cleaved by ACE to create Ang II, the main effector in the RAAS. The synthesis and secretion of aldosterone, another effector molecule in the RAAS, are stimulated by Ang II through the AT1 receptor in the adrenal cortex [28].

Although the rats in the HSD group consumed more calories due to the addition of doenjang, BW was not significantly different between the three groups. Liver weight, GOT, and GPT levels did not significantly between the groups; however, the liver/BW ratio was significantly lower in the HSD group than in the HS group. The liver weight in 2.3% NaCl group was significantly bigger than in commercial diet (0.3% NaCl) [29]. However, hepatic weight in the Japanese soybean paste (miso) diet group was decreased than in the NaCl diet group [29]. The initial stage of hypertension was not associated with overall renal failure, a small number of glomeruli, or glomerular hypertension [29]. Salt intake for eight weeks did not cause hepatic or renal injury, likely because of the short-term ingestion. The metabolism of doenjang is thought to be slower compared to the response to high-salt diet and decreases the risk of development and progression of hypertension and kidney damage.

Salt intake is an essential factor in controlling urination and water intake. Higher salt consumption increases urinary output and water intake to comparable levels. Increases in serum sodium and serum osmotic pressure stimulate thirst and antidiuretic hormones, causing increased fluid intake, decreased serum osmotic pressure, and increased urine volume [30].

A high level of sodium intake is reportedly associated with increased blood pressure. In the present study, the elevated blood pressure observed in rats fed a high-salt diet was consistent with previous studies [30,31]. Changes in systolic blood pressure were studied in spontaneously hypertensive rats fed a doenjang for nine weeks. Doenjang fermented with *Monascus* koji significantly decreased systolic blood pressure compared with a commercial diet [32]. Watanabe et al. showed that miso reduced blood pressure in SD and Dahl rats despite high salt content. Stroke-prone spontaneously hypertensive rats fed a miso diet showed decreased blood pressure and increased life survival compared with rats fed a high-salt diet containing 2.5% NaCl [33]. Ingestion of soybean paste, including doenjang and miso, is considered a significant source of dietary salt that does not increase blood pressure. Triple injections of His-His-Leu isolated peptides from doenjang lowered systolic blood pressure in spontaneously hypertensive rats [34]. In previous studies, doenjang extract exhibited ACE inhibitory activity in vitro [25]. When consuming fermented doenjang, dipeptides such as arginine-proline, which can produce an ACE inhibitory effect, are also ingested. Dipeptides and other peptide substances are known to prevent high blood pressure or have blood pressure depressant abilities [24].

The relationship between estimated potassium excretion and blood pressure is consistent with a systolic pressure of 0.65 mmHg per gram of potassium and a decrease in diastolic pressure of 0.42 mmHg per gram [35]. Sodium and potassium excretion, which is estimated as surrogate marker for ingestion, and blood pressure records in adults showed a non-linear correlation. If sodium excretion remains constant and potassium excretion is high, blood pressure decreases [35]. The sodium excreted in urine was not significantly different in the HS and HSD groups; however, the potassium excreted in urine was significantly elevated in the HSD group than in the HS group. The difference in ion excretion may also have contributed to lower blood pressure in rats in the HSD group.

In conventional theory, the typical reaction to the increase in salt intake is the suppression of circulating RAA hormones. Circulating renin activity and aldosterone concentrations were actively suppressed by a high-sodium diet [36]. Changes in arterial pressure were not observed with the high-sodium diet, although levels of Ang II and aldosterone were lessened in the 4% and 8% NaCl groups [19]. However, Wang et al. propose that renal RAAS action is independent or Versa of plasma RAA level at high salt intake, and improper activation of the RAAS in the kidney may contribute directly to hypertension and kidney damage [37]. Although the high-salt diet did not inhibit RAA level, doenjang lowered serum levels of renin and aldosterone in the present study.

A high-salt diet increased the mRNA expression of RAAS components in the kidney cortex. AT1 receptor and ACE mRNA expression in the kidney cortex was higher in the high-salt diet than in normal-salt diet groups. The RAAS in the kidney was inappropriately enhanced by high salt intake in SD rats [37]. High salt intake exacerbated blood pressure elevation during the development of hypertension in spontaneously hypertensive rats [38]. The increased salt content in the diet stimulates glomerular oxidative stress, which leads to the AT1 receptor, MR, and ACE up-regulation, subsequently causing hyperbole in sodium transporters, and providing to sodium retention and hypertension [39,40]. A high-salt diet could cause kidney damage through oxidative stress and modulation in the ACE/ACE2 ratio [41].

Electrogenic cationic pumps, as well as NKA α1, NCX, NBC, and NHE3, are essential for transcellular movement of water and ions in osmoregulatory epithelia. Increasing Ang II concentration decreased NKA activity in eel enterocytes; however, NKA activity was not affected by concentration of Ang II in saline adaptation [42]. The abundance of NHE3 and NKAα1 in the kidney cortex and medulla was not changed in SD rats fed a 4% NaCl diet [43]. A chronic high-salt diet increased renal NCX1 mRNA expression in Wistar–Kyoto rats [44].

Regarding the regulation of the sodium–potassium pump, Ang II activates PKCβ at low concentrations. As a result, the NKAα1 subunit is phosphorylated, and NKA is incorporated into the plasma membrane. Aldosterone is reported to stimulate the migration of the NKAα1 subunit from the intracellular compartment to the basal side membrane surface of the distal nephron. NBC-1 plays a vital role in regulating renal acid-alkaline balance, maintaining pH in the blood and the cells, and regulating sodium transport through NHE-3 in the proximal tubule. The regulations of NHE3 and NBC-1 have standard features due to the interrelated functions of the two ion transporters. The isolation of individual levels in function regulation is severe because many intracellular factors are involved in modulating the transporter function at different levels and may act as universal regulators of several transporters [45].

In this study, we confirmed that doenjang affects the RAA mechanism. However, doenjang containing high salt does not have a direct effect on blood pressure increase in a healthy animal model. The results from the present study indicate the importance of the microbial community in doenjang. The majority of microorganisms in doenjang are *Bacillus paralicheniformis*, which is tolerant to 10% NaCl [46]. Furthermore, probiotic microorganisms, including *Bacillus* strains, may have exerted health benefits. The consumption of probiotics may improve blood pressure control. Probiotics can be used as a potential supplement for future interventions to prevent high blood pressure or improve blood pressure control [47]. Although the mechanism is different from that of the RAAS, probiotics are expected to contribute to blood pressure control.

The primary sources of dietary sodium for the Korean people are kimchi, salt, soy sauce, and soybean paste [16]. Because doenjang significantly contributes to salt intake in Koreans, the focus in the present study was on the effects of salt contained in doenjang on blood pressure. The consumption of a large amount of traditional, salty fermented food was not found to be a direct cause of hypertension. For example, the high consumption of kimchi was not associated with an increased risk of hypertension in Korean adults [48]. The results from the present study showed that the intake of doenjang in normal healthy animals might improve blood pressure.

5. Conclusions

Our study demonstrated that doenjang with high-salt content improved blood pressure in SD rats. The intake of doenjang increased potassium excretion in feces and urine compared to high salt intake. Furthermore, doenjang decreased renin and aldosterone levels in serum and the expression of the AT1 receptor, ACE, and MR in the kidney cortex compared to high salt intake. These results indicate that eating traditional salty fermented food is not a direct cause of hypertension, and the intake of doenjang in normal healthy animals improved blood pressure.

Author Contributions: Formal analysis, E.-G.M.; Funding acquisition, Y.-S.C.; Methodology, E.-G.M.; Project administration, Y.-S.C.; Supervision, Y.-S.C.; Validation, Y.-S.C.; Writing—original draft, E.-G.M.; Writing-Review & Editing for revision, J.E.P.

Funding: This research was supported by Basic Science Research Program through the National Research Foundation of Korea (NRF) funded by the Ministry of Education (2015R1D1A3A01015869).

Conflicts of Interest: The authors declare no conflict of interest.

References

1. Lee, D.H.; Kim, M.J.; Park, S.H.; Song, E.J.; Nam, Y.D.; Ahn, J.; Jang, Y.J.; Ha, T.Y.; Jung, C.H. Bioavailability of Isoflavone Metabolites After Korean Fermented Soybean Paste (Doenjang) Ingestion in Estrogen-Deficient Rats. *J. Food Sci.* **2018**, *83*, 2212–2221. [CrossRef] [PubMed]
2. Park, S.K.; Seo, K.I. Quality Assessment of Commercial Doenjang Prepared by Traditional Method. *J. Korean Soc. Food Sci. Nutr.* **2000**, *29*, 211–217.
3. Park, N.Y.; Rico, C.W.; Lee, S.C.; Kang, M.Y. Comparative effects of doenjang prepared from soybean and brown rice on the body weight and lipid metabolism in high fat-fed mice. *J. Clin. Biochem. Nutr.* **2012**, *51*, 235–240. [CrossRef] [PubMed]
4. Kwak, C.S.; Park, S.C.; Song, K.Y. Doenjang, a fermented soybean paste, decreased visceral fat accumulation and adipocyte size in rats fed with high fat diet more effectively than nonfermented soybeans. *J. Med. Food* **2012**, *15*, 1–9. [CrossRef] [PubMed]

5. Cha, Y.S.; Yang, J.A.; Back, H.I.; Kim, S.R.; Kim, M.G.; Jung, S.J.; Song, W.O.; Chae, S.W. Visceral fat and body weight are reduced in overweight adults by the supplementation of Doenjang, a fermented soybean paste. *Nut. Res. Pract.* **2012**, *6*, 520–526. [CrossRef] [PubMed]
6. Lee, M.; Chae, S.; Cha, Y.; Park, Y. Supplementation of Korean fermented soy paste doenjang reduces visceral fat in overweight subjects with mutant uncoupling protein-1 allele. *Nutr. Res.* **2012**, *32*, 8–14. [CrossRef] [PubMed]
7. Cha, Y.S.; Park, Y.; Lee, M.; Chae, S.W.; Park, K.; Kim, Y.; Lee, H.S. Doenjang, a Korean fermented soy food, exerts antiobesity and antioxidative activities in overweight subjects with the PPAR-gamma2 C1431T polymorphism: 12-week, double-blind randomized clinical trial. *J. Med. Food* **2014**, *17*, 119–127. [CrossRef] [PubMed]
8. Chung, S.I.; Rico, C.W.; Kang, M.Y. Comparative study on the hypoglycemic and antioxidative effects of fermented paste (doenjang) prepared from soybean and brown rice mixed with rice bran or red ginseng marc in mice fed with high fat diet. *Nutrients* **2014**, *6*, 4610–4624. [CrossRef] [PubMed]
9. Kim, M.S.; Kim, B.; Park, H.; Ji, Y.; Holzapfel, W.; Kim, D.Y.; Hyun, C.K. Long-term fermented soybean paste improves metabolic parameters associated with non-alcoholic fatty liver disease and insulin resistance in high-fat diet-induced obese mice. *Biochem. Biophys. Res. Commun.* **2018**, *495*, 1744–1751. [CrossRef] [PubMed]
10. Jung, K.O.; Park, S.Y.; Park, K.Y. Longer aging time increases the anticancer and antimetastatic properties of doenjang. *Nutrition* **2006**, *22*, 539–545. [CrossRef] [PubMed]
11. Jeong, J.K.; Chang, H.K.; Park, K.Y. Doenjang prepared with mixed starter cultures attenuates azoxymethane and dextran sulfate sodium-induced colitis-associated colon carcinogenesis in mice. *J. Carcinog.* **2014**, *13*, 1–9.
12. Nam, Y.R.; Won, S.B.; Chung, Y.S.; Kwak, C.S.; Kwon, Y.H. Inhibitory effects of Doenjang, Korean traditional fermented soybean paste, on oxidative stress and inflammation in adipose tissue of mice fed a high-fat diet. *Nutr. Res. Pract.* **2015**, *9*, 235–241. [CrossRef] [PubMed]
13. Lee, J.H.; Paek, S.H.; Shin, H.W.; Lee, S.Y.; Moon, B.S.; Park, J.E.; Lim, G.D.; Kim, C.Y.; Heo, Y. Effect of fermented soybean products intake on the overall immune safety and function in mice. *J. Vet. Sci.* **2017**, *18*, 25–32. [CrossRef] [PubMed]
14. Jang, S.E.; Kim, K.A.; Han, M.J.; Kim, D.H. Doenjang, a fermented Korean soybean paste, inhibits lipopolysaccharide production of gut microbiota in mice. *J. Med. Food* **2014**, *17*, 67–75. [CrossRef] [PubMed]
15. Park, S.M.; Oh, J.; Kim, J.E.; Kim, J.S. Effect of Drying Conditions on Nutritional Quality and In Vitro Antioxidant Activity of Traditional Doenjang. *Prev. Nutr. Food Sci.* **2018**, *23*, 144–151. [CrossRef] [PubMed]
16. Lee, H.S.; Duffey, K.J.; Popkin, B.M. Sodium and potassium intake patterns and trends in South Korea. *J. Hum. Hypertens.* **2013**, *27*, 298–303. [CrossRef] [PubMed]
17. Härtl, G. WHO issues new guidance on dietary salt and potassium. *Cent. Eur. J. Public Health* **2013**, *21*, 16.
18. Nakamura, K.; Miyoshi, T. The optimal amount of salt intake. *Hypertens. Res.* **2019**, *42*, 752–753. [CrossRef] [PubMed]
19. Crestani, S.; Gasparotto Junior, A.; Marques, M.C.; Sullivan, J.C.; Webb, R.C.; da Silva-Santos, J.E. Enhanced angiotensin-converting enzyme activity and systemic reactivity to angiotensin II in normotensive rats exposed to a high-sodium diet. *Vasc. Pharmacol.* **2014**, *60*, 67–74. [CrossRef] [PubMed]
20. Kobori, H.; Nishiyama, A.; Abe, Y.; Navar, L.G. Enhancement of intrarenal angiotensinogen in Dahl salt-sensitive rats on high salt diet. *Hypertension* **2003**, *41*, 592–597. [CrossRef] [PubMed]
21. Susic, D.; Frohlich, E.D.; Kobori, H.; Shao, W.; Seth, D.; Navar, L.G. Salt-induced renal injury in SHRs is mediated by AT1 receptor activation. *J. Hypertens.* **2011**, *29*, 716–723. [CrossRef] [PubMed]
22. Korkmaz, D. Precipitation titration:"Determination of Chloride by the Mohr Method". *Methods* **2001**, *2*, 1–6.
23. Kim, Y.S.; Kim, M.C.; Kwon, S.W.; Kim, S.J.; Park, I.C.; Ka, J.O.; Weon, H.Y. Analyses of bacterial communities in meju, a Korean traditional fermented soybean bricks, by cultivation-based and pyrosequencing methods. *J. Microbiol.* **2011**, *49*, 340–348. [CrossRef] [PubMed]
24. Shin, D.; Jeong, D. Korean traditional fermented soybean products: Jang. *J. Ethnic Foods* **2015**, *2*, 2–7. [CrossRef]
25. Mun, E.G.; Kim, B.; Kim, E.Y.; Lee, H.J.; Kim, Y.; Park, Y.; Cha, Y.S. Research Trend in Traditional Fermented Foods Focused on Health Functional Evaluation. *J. Korean Soc. Food Sci. Nutr.* **2018**, *47*, 373–386. [CrossRef]

26. Daliri, E.B.; Ofosu, F.K.; Chelliah, R.; Park, M.H.; Kim, J.H.; Oh, D.H. Development of a Soy Protein Hydrolysate with an Antihypertensive Effect. *Int. J. Mol. Sci.* **2019**, *20*, 1496. [CrossRef] [PubMed]
27. Kalupahana, N.S.; Moustaid-Moussa, N. The renin-angiotensin system: A link between obesity, inflammation and insulin resistance. *Obes. Rev.* **2012**, *13*, 136–149. [CrossRef] [PubMed]
28. Munoz-Durango, N.; Fuentes, C.A.; Castillo, A.E.; Gonzalez-Gomez, L.M.; Vecchiola, A.; Fardella, C.E.; Kalergis, A.M. Role of the Renin-Angiotensin-Aldosterone System beyond Blood Pressure Regulation: Molecular and Cellular Mechanisms Involved in End-Organ Damage during Arterial Hypertension. *Int. J. Mol. Sci.* **2016**, *17*, 797. [CrossRef] [PubMed]
29. Watanabe, H.; Kashimoto, N.; Kajimura, J.; Kamiya, K. A miso (Japanese soybean paste) diet conferred greater protection against hypertension than a sodium chloride diet in Dahl salt-sensitive rats. *Hypertens. Res.* **2006**, *29*, 731–738. [CrossRef] [PubMed]
30. Oloyo, A.K.; Imaga, N.O.A.; Fatope, Y.; Sofola, O.A. Sex differences in cardiac and renal responses to a high salt diet in Sprague-Dawley rats. *Heliyon* **2019**, *5*, e01665. [CrossRef] [PubMed]
31. Mun, E.G.; Sohn, H.S.; Kim, M.S.; Cha, Y.S. Antihypertensive effect of Ganjang (traditional Korean soy sauce) on Sprague-Dawley Rats. *Nutr. Res. Prac.* **2017**, *11*, 388–395. [CrossRef] [PubMed]
32. Rhyu, M.R.; Kim, E.Y.; Han, J.S. Antihypertensive effect of the soybean paste fermented with the fungus Monascus. *Int. J. Food Sci. Technol.* **2002**, *37*, 585–588. [CrossRef]
33. Watanabe, H.; Sasatani, M.; Doi, T.; Masaki, T.; Satoh, K.; Yoshizumi, M. Protective Effects of Japanese Soybean Paste (Miso) on Stroke in Stroke-Prone Spontaneously Hypertensive Rats (SHRSP). *Am. J. Hypertens.* **2017**, *31*, 43–47. [CrossRef] [PubMed]
34. Shin, Z.I.; Yu, R.; Park, S.A.; Chung, D.K.; Ahn, C.W.; Nam, H.S.; Kim, K.S.; Lee, H.J. His-His-Leu, an angiotensin I converting enzyme inhibitory peptide derived from Korean soybean paste, exerts antihypertensive activity in vivo. *J. Agric. Food Chem.* **2001**, *49*, 3004–3009. [CrossRef] [PubMed]
35. Mente, A.; O'Donnell, M.J.; Rangarajan, S.; McQueen, M.J.; Poirier, P.; Wielgosz, A.; Morrison, H.; Li, W.; Wang, X.; Di, C.; et al. Association of urinary sodium and potassium excretion with blood pressure. *N. Engl. J. Med.* **2014**, *371*, 601–611. [CrossRef] [PubMed]
36. Engeli, S.; Boschmann, M.; Frings, P.; Beck, L.; Janke, J.; Titze, J.; Luft, F.C.; Heer, M.; Jordan, J. Influence of salt intake on renin-angiotensin and natriuretic peptide system genes in human adipose tissue. *Hypertension* **2006**, *48*, 1103–1108. [CrossRef] [PubMed]
37. Wang, Y.; Xie, B.Q.; Gao, W.H.; Yan, D.Y.; Zheng, W.L.; Lv, Y.B.; Cao, Y.M.; Hu, J.W.; Yuan, Z.Y.; Mu, J.J. Effects of Renin Angiotensin System Inhibitors on Renal Expression of Renalase in Sprague-Dawley Rats Fed with High Salt Diet. *Kidney Blood Press. Res.* **2015**, *40*, 605–613. [CrossRef] [PubMed]
38. Koga, Y.; Hirooka, Y.; Araki, S.; Nozoe, M.; Kishi, T.; Sunagawa, K. High salt intake enhances blood pressure increase during development of hypertension via oxidative stress in rostral ventrolateral medulla of spontaneously hypertensive rats. *Hypertens. Res.* **2008**, *31*, 2075–2083. [CrossRef] [PubMed]
39. Banday, A.A.; Lokhandwala, M.F. Oxidative stress-induced renal angiotensin AT1 receptor upregulation causes increased stimulation of sodium transporters and hypertension. *Am. J. Physiol. Renal Physiol.* **2008**, *295*, F698–F706. [CrossRef] [PubMed]
40. Funder, J.W. Mineralocorticoid receptor activation and oxidative stress. *Hypertension* **2007**, *50*, 840–841. [CrossRef] [PubMed]
41. Bernardi, S.; Toffoli, B.; Zennaro, C.; Tikellis, C.; Monticone, S.; Losurdo, P.; Bellini, G.; Thomas, M.C.; Fallo, F.; Veglio, F.; et al. High-salt diet increases glomerular ACE/ACE2 ratio leading to oxidative stress and kidney damage. *Nephrol. Dial. Transplant.* **2012**, *27*, 1793–1800. [CrossRef] [PubMed]
42. Marsigliante, S.; Muscella, A.; Barker, S.; Storelli, C. Angiotensin II modulates the activity of the Na+/K+ATPase in eel kidney. *J. Endocrinol.* **2000**, *165*, 147–156. [CrossRef] [PubMed]
43. Yang, L.E.; Sandberg, M.B.; Can, A.D.; Pihakaski-Maunsbach, K.; McDonough, A.A. Effects of dietary salt on renal Na+ transporter subcellular distribution, abundance, and phosphorylation status. *Am. J. Physiol. Renal Physiol.* **2008**, *295*, F1003–F1016. [CrossRef] [PubMed]
44. Yatabe, M.S.; Yatabe, J.; Takano, K.; Murakami, Y.; Sakuta, R.; Abe, S.; Sanada, H.; Kimura, J.; Watanabe, T. Effects of a high-sodium diet on renal tubule Ca2+ transporter and claudin expression in Wistar-Kyoto rats. *BMC Nephrol.* **2012**, *13*, 160. [CrossRef] [PubMed]
45. Abramicheva, P.A.; Smirnova, O.V. Role of hormones in regulating sodium transporters in the kidney: Modulation of phosphorylation, traffic, and expression. *Hum. Physiol.* **2017**, *43*, 474–487. [CrossRef]

46. Dunlap, C.A.; Kwon, S.W.; Rooney, A.P.; Kim, S.J. Bacillus paralicheniformis sp. nov., isolated from fermented soybean paste. *Int. J. Syst. Evol. Microbiol.* **2015**, *65*, 3487–3492. [CrossRef] [PubMed]
47. Khalesi, S.; Sun, J.; Buys, N.; Jayasinghe, R. Effect of probiotics on blood pressure: A systematic review and meta-analysis of randomized, controlled trials. *Hypertension* **2014**, *64*, 897–903. [CrossRef] [PubMed]
48. Song, H.J.; Park, S.J.; Jang, D.J.; Kwon, D.Y.; Lee, H.J. High consumption of salt-fermented vegetables and hypertension risk in adults: A 12-year follow-up study. *Asia Pac. J. Clin. Nutr.* **2017**, *26*, 698–707. [PubMed]

MDPI
St. Alban-Anlage 66
4052 Basel
Switzerland
Tel. +41 61 683 77 34
Fax +41 61 302 89 18
www.mdpi.com

Nutrients Editorial Office
E-mail: nutrients@mdpi.com
www.mdpi.com/journal/nutrients

www.ingramcontent.com/pod-product-compliance
Lightning Source LLC
Chambersburg PA
CBHW062047220926
51850CB00010B/2683
* 9 7 8 3 0 3 9 3 6 4 6 5 7 *